Design Research Foundations

Series Editors

The goal of the series is to provide a platform for publishing state of the art research on foundational issues in design and its applications in industry and society. Suitable topics range from methodological issues in design research to philosophical reflections on the specificities of design rather than actual design work or empirical cases only. The definition of design behind the series is inclusive. In terms of disciplines, it ranges from engineering to architecture. In terms of design work, it ranges from conceptual issues in design through design experiments and prototypes to evaluative studies of design and its foundations.

Proposals should include:

A proposal form, as can be found on this page
A short synopsis of the work or the introduction chapter
The proposed Table of Contents
The CV of the lead author(s)
If available: one sample chapter

We aim to make a first decision within 1 month of submission. In case of a positive first decision the work will be provisionally contracted: the final decision about publication will depend upon the result of the anonymous peer review of the complete manuscript. The series editors aim to have the complete work peer-reviewed within 3 months of submission.

The series discourages the submission of manuscripts that contain reprints of previous published material and/or manuscripts that are below 150 pages/75,000 words.

For inquiries and submission of proposals authors can contact the series editors, Pieter Vermaas via: p.e.vermaas@tudelft.nl or Ilpo Koskinen via: ilpo.koskinen@unsw.edu.au

More information about this series at http://www.springer.com/series/13775

Peter Gall Krogh • Ilpo Koskinen

Drifting by Intention

Four Epistemic Traditions from within Constructive Design Research

Springer

Peter Gall Krogh
Department of Engineering
Socio-Technical design
Aarhus University
Aarhus, Denmark

Ilpo Koskinen
Design Next
University of New South Wales
Sydney, NSW, Australia

ISSN 2366-4622 ISSN 2366-4630 (electronic)
Design Research Foundations
ISBN 978-3-030-37895-0 ISBN 978-3-030-37896-7 (eBook)
https://doi.org/10.1007/978-3-030-37896-7

This Springer imprint is published by the registered company Springer Nature Switzerland AG
The registered company address is: Gewerbestrasse 11, 6330 Cham, Switzerland

Preface

The term constructive design research captures what happens to design when it shifts from building products, services, jewelry, interactive systems, textiles, fashion objects, etc. to creating knowledge. This book takes a look at what happens in constructive design research from within and tells how we understand "drifting" as a native and key characteristic of this form of research. We believe our analysis is also useful for design practitioners.

For us, drifting is what every designer does in every project – and what every design researcher does in every research project. By drifting, we mean those actions that take design away from its original brief or question and lead to a result that was not anticipated in the beginning. We talk about drifting as a conscious process that happens intentionally and depends on certain epistemology. One designer drifts in the studio, another in critiques, a third in the world of simulations, a fourth through randomization, and a fifth using many different tactics. Every designer – and design researcher – drifts intentionally and methodically. This process can sometimes lead to radical changes, but it does not have to. In fact, we argue that design literature needs a more modest and nuanced description of design that does not reduce design into a dialogue between problems and solutions that spur reframing, blue-sky thinking, radical and disruptive innovations, or counterintuitive findings. It may be, but these heroic moments are rare and, as we try to show, should not be turned into a standard expectation or a norm. We need a more flexible way to understand those shifts that characterize design and design research, and drifting is our word for them.

Before this gets too philosophical, it is perhaps a good idea to give an example.

Odo Fioravanti is an Italian designer who is best known for his chairs, so many on the market by now that his friends call him Mr. Chairman. In December 2016, he gave a talk about his work in Hong Kong, describing two-dozen design pieces he has done and detailing the usual material innovations behind them. After his talk, a member of the audience challenged him by asking whether he is aware of the mathematical and scientific tools available to him at all. Rather than being agitated by the question, which was meant to show the superiority of science and engineering vis-à-vis designers, Fioravanti raised his head, gave a friendly smile to the audience, and responded calmly: "Well, I studied mechanical engineer and I know its methods.

Of course I can do finite elements, for instance. However, experience is much more important."

Fioravente's response well captured two cultures that characterize design. For a master of his trade like him, experience is one way to design. While the methodic path would lead to modeling chairs mathematically, the experience-based path would lead him to a close study of materials, ergonomics, joinery, form language, molding, and attachments in existing chairs. He does not say that science is unimportant, but he acknowledges other ways of knowing. For him, the true question is how knowledge works in design practice, how it is generated, and how it can be recovered for the same practice. These are the skills any designer deals with in daily work – the important reason why studio-based work in both education and research is key to design and constructive forms of design research. There is knowledge in practice, even though it is different from those scientific discourses that usually dominate debates in design research and find their opponents from practitioners.

The exchange in Hong Kong is wanting in one respect, however. It captures two ways of understanding knowledge. This book argues in contrast that there are at least two more ways to understand knowledge, its implications to drifting, and how these implications shape the shift from design practice to a knowledge-based discipline.

Our way of analyzing this shift is epistemological. In philosophy, epistemology means the study of knowledge. It probes what are the conditions under which we can trust some statement and take it to be true (see, for instance, Feldman 2003). As it is easy to imagine, the answers to this age-old question have been far and between. Philosophers like Plato have postulated a world of independent, invisible ideas that are reflected in reality. This approach, well-rooted in design's concept of forms as signs for use and meaning, turns research into a search of these ideas. Rationalists like Rene Descartes have denied the possibility of knowing the world and trust our logical abilities – the only thing we know for sure is that it is us who are thinking, which means that knowledge must be sought from our own thinking and not from the objects that we surround our self with. Empirical philosophers like John Locke and David Hume and logical empiricists like Otto Neurath and Rudolf Carnap on their part see senses as a source of knowledge (Locke famously saw humans as *tabula rasa*) and usually analyze how we can know whether sense datum can be trusted. For example, how do we know whether what we see is not a dream or a hallucination? In an attempt to reconcile these two, Immanuel Kant proposed that knowledge lies in universal categories of the mind, a view that later gave birth to both phenomenology and cognitive science. Phenomenologists like Martin Heidegger and Hans-Georg Gadamer go beyond epistemology to study how is it possible to know anything at all, while others follow Ludwig Wittgenstein's late philosophy and analyze how knowledge depends on our uses of language and a "form of life" behind language.

The most recent versions of epistemology, however, have tended to be historical and empirical. They are, for instance, studies of how science sometimes begins to treat some facts as paradigmatic and how these untouchable truths are sometimes revolutionized (Kuhn 1962) or how facts are manufactured in laboratories and research instituted with ethnographic tools (Latour and Woolgar 1979; Lynch 1993; Collins 1985). Many of these arguments have followers in design communities as well – all the way from the implicit Neoplatonism of the Bauhaus (as Otl Aicher 2009 describes the legacy of the Bauhaus) to Johan Redström's (2017) recent Kuhnian-inspired interpretation of design.

Our preferred analysis would have started from existing epistemological writing in design and engineering, but we did not find a lot (e.g., see Schmid 2018). Those few sources we managed to find (see Mitcham 1986 and van Fraassen et al. 1991) main point out that epistemology in the world-making disciplines is heavily embedded in culture and society at large. As Mitcham, for example, points out, since everything man-made, computers are embodiments of human values and as such give a material form to ethical problems that should be studied epistemologically. His analysis is useful and well in line with how we see design from within, but he does not, however, give us a method for explicating these values.

Given the paucity of literature, our take on epistemology has been inspired by the empirical studies of science referred to above, but our method is explication rather than ethnographic. We study some of the key texts in literature and explicate their understanding of knowledge – that is, we try to capture their implicit epistemology. Then we turn our attention to constructive design research as such, asking how these interpretations work in doctoral level research. We shall not argue that epistemology somehow determines how design research is done (or, even worse, how it should be done) but argue that seeing it as an epistemic practice helps to understand it better. An epistemology from within, as we call it, tells researchers how to treat previous research and how to select methods and methodologies. It also tells them the nuts and bolts of research: which arguments are strong enough to change the direction of research, how to experiment, how to evaluate their findings, and how to communicate the outcomes. It also gives a tool that helps to open up implicit universal assumptions in disciplines of design and clarifies the relationship between some of the debates that have shaped up first design research and then design itself over the last two and a half decades.

The main contribution of this book is the concept of drifting and the insights it provides for a better understanding of how design researchers approach their craft. We believe that if we gain a better understanding of epistemology in design as something that comes from within the discipline, it can become a useful addition to the discipline. Some will argue that design has become a research discipline on its own rights as it offers particular ways of approaching research and development challenges and deliver knowledge that is particularly relevant and significant to its own community. Others will rightfully state that design and constructive design research borrows heavily, theoretically, from other disciplines. However, this accounts for most other disciplines too. Being a research discipline on its own rights is not a matter of crossing a well-defined finish line but a matter of maturity. Looking to the material, most prominently the corpus of PhD dissertations that we analyze, constructive design research by now has well-established publication fora and delivered unique valued and respected contributions to research literature in management, engineering, policy-making, human computer interaction, and many others. This, at least, tells of a maturing field.

Our broader aim is to open up a new way in which designers can see their work. Most things we say are already known; what we provide is a coherent interpretation of arguments that exist in literature and a vocabulary for analyzing them. The aim is clarification; the method is an explication of arguments that already exists. If this clarification helps design researchers to understand better the foundations of their work, it will, we believe, provide them the freedom they need to navigate our increasingly academic discipline.

Aarhus, Denmark
Sydney, Australia
June 2019

Peter Gall Krogh
Ilpo Koskinen

Acknowledgment

This book and its thoughts have been in the making for quite a few years. During these years, the ideas changed and matured. The way of maturation as can be traced in our evolving presentations and coursework was carried out in diverse design programs. Iterating ideas and versions of models, exposing them for discussion, has, to us, been an essential part of developing thinking behind this book. There are many people to thank for their challenging discussions and for their generosity to allow yet half-baked thoughts live in discussions. It is hard to direct well-measured amounts of thanks to them. Not only experiences from presentations and academic discussions but also casual hallway chats have at times led to thoughts later to be discussed between the authors; some have found their way into the book. We are grateful to all these people. However, we also recognize that the efforts on this matter are prone to error. We apologize in advance to anyone who would have desired more credit. Having stated this, there is a set of people we would like specifically to extend our thanks to beyond academic credits.

In 2011, when Peter Gall Krogh was professor at Aarhus School of Architecture in Jutland, Denmark, he initiated a series of 3-week long courses with three Danish colleagues of his. These courses were planned at helping PhD students to comfortably pursue new knowledge during their doctoral studies by applying their design competences. We are grateful to the management of Aarhus School of Architecture and Kolding School of Design for supporting these courses, and for the generous sponsoring by the Danish Ministry of Culture. The courses were developed in collaboration with Thomas Markussen, Anne Louise Bang who at that time was in employed at Kolding School of Design, and Martin Ludvigsen, one of the most talented design researchers we have ever seen who was also a colleague at Aarhus School of Architecture. As authors of the book, we are grateful to their valuable contributions to the early thoughts, also expressed in coauthored papers, that are key to Chaps. 4, 5, and 6. Without these work and contributions, the book would have been a very different and less operational.

Above all, we are indebted to the more than 40 doctoral individual students from all over Europe. They had the courage to expose their research to early versions of models and methods presented in this book. They believed that their research might improve from taking part in the courses. Their uses and reappropriations of our early versions of ideas helped us rephrase the models and concepts presented in the book.

In 2010, Ilpo Koskinen was visiting professor at Aarhus School of Architecture while writing parts of the manuscript for *Design Research Through Practice: From the Lab, Field, and Showroom*. Some of the formative ideas of the current book had their origins in that visit. We would also like to thank two of the coauthors of *Design Research Through Practice*, Johan Redström and Stephan Wensveen, who contributed lectures to the doctoral courses in Jutland and later discussed ideas of the book with the first author in detail.

During the period of writing, earlier versions of its ideas have been presented and discussed on a range of occasions. We would like specifically to *thank* the School of Design of the Hong Kong Polytechnic University for generously sponsoring lectures and visits and for allowing the coauthors to meet to advance the manuscript. In the same vein, we like to thank Mette Agger Eriksen and Pelle Ehn for constructive discussions in connection to talks at Malmö University. Our thanks also have to go to Alessandro Deserti and Ezio Manzini at Politecnico di Milano and Bill Gaver and his research group at the Interaction Design Studio at Goldsmiths, University of London.

Special thanks are extended to Qian Xiaobo and Chen Xiangjie at Jiangnan University School of Design in China for inviting and generously hosting Peter Gall Krogh as visiting professor. Their generosity has provided Krogh an opportunity to present and discuss the concepts and models of the book with PhD students from a wide range of locations in China and also local master and PhD students in Jiangsu province. The insights gained from these encounters have helped us to tidy up the models. They have also helped us to better articulate their value. Without the hospitality of the Jiangnan University and the willingness of the students to engage in the material, the applicability of the thoughts would have been less clear.

Last but not least, we would like to thank the Socio-technical Design Group in the Department of Engineering, Aarhus University, Minna Pakkanen, Majken Kirkegaard Rasmussen, Kasper Heiselberg, Sofie Kinch, and Troy Nachtigal, for discussions and valuable comments on the drafts of the manuscript. In addition to this, we like to thank Majken and Minna together with Christine Tsin for getting the graphics right. Finally, several cohorts of students in the master's program of IT Product Development at Aarhus University have been discussing the ideas of the book over their studies. Such forgiving and invigorating discussions are what makes academic life fun and rewarding! *Thanks*!

Aarhus and Sydney
February 2020

Contents

Chapter 1
Drifting

In this book, we deal with the epistemology of design. Epistemology is one of the grand terms of philosophy, where it means discourse about knowledge — usually about certainty, whether we can trust our senses, thoughts, and other pieces of knowledge. For us, two design researchers, the concept is more specific. We deal with epistemology in one particular context, constructive design research, in which design artefacts are vehicles of knowledge creation. In our work we have identified four epistemic traditions in constructive design research: (1) *experiential;* (2) *methodic;* (3) *programmatic;* and (4) *dialectic*. These are described in Chap. 3. Our aim is not to contribute to philosophy; our aim is to clarify how knowledge works in constructive design research. Rather than tightening the bridle this book is our attempt to maximize the freedom of research as it happens in constructive design research. The descriptions and methodologies provided in this book is our attempt to give research legitimacy to a cherished design practice we call drifting — however, drifting by intention.

Drifting is typical to design, and cannot be avoided in it. We believe understanding drifting is crucial for understanding design and constructive design research. The concept of drifting also resonates with modest approaches to design rather than what we regard as currently overdramatized concepts like reframing, radical change, design driven innovation and so on.[1] While there are cases in which designers have dramatically redefined product categories, these are exceptions in an industry that mostly creates small improvements, and which equally often recycles old forms for commercial purposes in manner described by the French philosopher Jean Baudrillard (1994) in his work on post-modernism. This is the reason for why we prefer to talk about drifting, which we see as an intentional activity: it is not

[1] We use the concept of innovation to cover two ways in which it is used in design. It sometimes means a field of activity, like in design schools. In engineering and business literature, it usually means novel inventions that are being commercialized. Under the first meaning, design bring new things to the world, but pay less attention to whether it creates value in the second sense. Our usage covers both cases.

P. G. Krogh, I. Koskinen, *Drifting by Intention*, Design Research Foundations,
https://doi.org/10.1007/978-3-030-37896-7_1

dramatic, and gives a picture of design as an occupation that indeed does change things and human relations, but usually through small, occasionally almost unnoticed changes. While carrying connotations of modesty, the concept of drifting does not rule out more dramatic shifts either, though. We believe it is better to see them as exceptions rather than as a rule. Design is much more than the peaking ideals, it is a dedicated profession also for all the mundane stuff we surround us with. A mundane de-dramatization also resonates with our Scandinavian heritage.

Throughout this book, we focus on ways in which drifting takes place when design becomes a research discipline. Our hypothesis builds on the idea of design accountability introduced in the writers' earlier work Ilpo Koskinen and Peter Gall Krogh (2015). As a research discipline, design sits on two chairs: design researchers want to make sure they are accountable — or at least intelligible — to practitioners, but in contrast to design practice, their work builds on knowledge and contributes to it. If design becomes an academic discipline, its relevance to practice suffers, we argued, unless it finds a way to create an academic base that is still understandable to practitioners (see Kees Dorst 2015). One corollary of this idea is that design practice sets limits to what can be done in design research. If we are correct, and design disciplines are different from each other, constructive design research should respect these differences as well. Product design and textile design are different, as are graphic design and interaction design. Similarities in creative process should not be taken as a proof against these differences, and this is the case of the identity of being a designer as well.

Our approach to these concerns builds on an analysis of various interpretations of knowledge in design. We claim that the way in which knowledge and practice work depends crucially on how we understand knowledge. Knowledge for us is more than scientific knowledge; it is also practical. Every cabinet maker knows that learning the craft takes years, and there are research techniques that maintain the skills that turn someone into a master of this trade. The concepts of knowledge in design research reflect this division, we believe. Hence, crucial to the undertaking of design research is our understanding knowledge and its implications. To put it on standard philosophical terms, when design becomes research, i.e. leaves the context of discovery and has to play the game of 'context of justification' (Karl Popper 2002: 7–8). With this shift, design changes to a ground defined by knowledge rather than practice alone. This has given us the main question of this book: how does drifting happen when designers have an imperative to produce knowledge and share it with research community?

In the past couple of decades there has been considerable steps to close a gap. Design departments in technical faculties and departments of universities have attempted to increasingly include art-based ways of working, and arts and crafts based design institutions and units have adopted a variety of research techniques, theories and methods to declare its relevance to other disciplines and society. To understand this shift, we have collected a corpus of PhD dissertations described in detail at the end of this book. A subset of these dissertations has served as analysis and reading material in a set of PhD courses offered to students with a background in arts and crafts based institutions while also inviting students from more mature

Fig. 1.1 PhD students using design skills to conduct research. Photo credit Lise Yde

disciplines. It is during such events that we have witnessed scenes as the below (Fig 1.1):

Vignette: PhD students in Denmark

20 PhD students of seven nationalities are sitting on the floor in a large seminar room. After months of mulling through piles of literature and theories they have been asked in drawing to individually sketch and in pairs discuss and explain, and if necessary to re-write their research plan and how they are planning this to build new knowledge. As the session progresses the initial hesitation and confusion is replaced by lively enthusiasm and as one of the students noted after the session: 'In the eagerness to live up to academic standards I've been reading so much that I had almost forgot the power I, as designer, have to explore and explain things when I draw and sketch!'

In our meeting with PhD students all over the world we have met many similar experiences and like-minded doctoral students in design; the preconception that in order to develop credible new knowledge I should first acquaint myself with science and theories from other disciplines before I can return to the practices and strengths of arts and crafts based design. Whereas doctoral students from university departments based on well-established technical or analytical research theories and methods initially have seen the training events as exotic glimpses into art based practices that may spur new ideas or a topic of study.

It is not the intent of this book to throw away sound and relevant knowledge of neighboring research disciplines and to replace it with imaginative creations of ink on paper. Neither is this book an attempt to turn classical sciences in to pseudo art. The purpose of this book is to point out how 'constructive design research' as it is termed by Koskinen and his colleagues (2011) may be regarded as a shared approach across institutionalized backgrounds. We claim that this is possible because the foundational challenge that what drives us all across these boundaries is: how do you design with the objective of producing knowledge?

Rather than looking at design as a topic and subject of research this book provides insights into how the act of designing is a vehicle for knowledge production and help researchers navigate and execute processes in which design is a research discipline. We have picked up our method from Eva Brandt and Thomas Binder (2007), who consulted three PhD dissertations to better understand how constructive design research is produced and documented, but we have extended their method from exemplars to a large corpus of over 60 dissertations. The three PhD dissertations they explored (Kristina Niedderer 2004; Adriaan Ianus 2005; and Tuuli Mattelmäki 2006) are included in our corpus too. While they '…map genealogy, compare interventions, and discuss how the outcomes of the various experiments become arguments in knowledge production,' we have adopted an approach in which the recurring observation of 'drift' in experiments, research interest and theory and so forth have shaped our analysis and findings. This has led us to:

1. Clarify constructive design research by analyzing four epistemological traditions at work in it;
2. Suggest a Knowledge-Relevance model (K-R) in Chap. 4 to articulate how constructive design research operates in relation to both knowledge production and relevance;
3. Analyze how hypothesis creation, design experimentation, evaluation and concept creation depend in epistemological traditions;
4. Analyze ways in which design experiments relate to each other when structured for the production of knowledge — five ways of drifting elaborated upon in Chap. 5.

1.1 Drifting in Profession and Research

Design is by now both a profession and a research discipline. In constructive design research, the designer and the researcher can be inseparable roles (Koskinen and Krogh 2015). Often, it is the same person who merges interests of knowledge production and concerns for producing relevant products. This is by no means a novel observation. Harold Nelson and Erik Stolterman (2003) point to design as an activity of producing 'Ultimate Particulars', Jan Pieter van Stappers and Elisa

Giaccardi (2017) show how design and research often carry different connotations and Sanders (2005) point to how design research is understood differently in academia and in industry.

The profession of design is often referred to as a 'solution oriented discipline.' Whereas it is true that design is concerned with delivering aesthetically attractive, useful, socially and commercially valuable solutions, it is often forgotten that design also is about questioning initial assumptions, convictions and theories in the pursuit of identifying the actual underlying challenges, concerns and questions. This happens in both professional practice and in research. Through producing sketches, artefacts, scenarios, mock-ups, service blueprints, screen dumps and video prototypes, the produced material meets the recurring sobering check: 'what question is this project an answer to?' As new knowledge evolves, the result points to findings that lead designers away from initial perceptions of the challenges and objective. This is drifting at its core and it is also a typical feature of design as a whole.

Drifting is found in most fields of research even though it is not always recognized as a proper scholarly practice. An illustrative example is a geologist who is trying to understand a ridge. He/she must study its history to understand how water, wind, and geological processes have shaped it. In this work, he/she makes notes and turns them into observations that help him/her to weave together a story that explains the observations. Throughout, he/she uses the best available methods to solve distinct problems as they come by. He/she may dig into libraries to find books that tell about the ridge. He/she may interview people who have lived by it. He/she may also interview other experts. He/she can take samples and analyze them with X-rays, spectrometry and radioactive carbon analysis. He/she also builds on theoretical knowledge geology and geomorphology have created over centuries.

In brief, he/she uses many types of methods and theories in putting together his/her story.

It is important to see that there are many types of knowledge in this story. In putting the story together, he/she makes errors and corrects them until both geologist and his/her peers are finally satisfied with it. His/her research drifts from one hunch to another; from one interpretation to another; from one line of inquiry to another. He/she carries on until his/her story is consistent and captures the essential elements that have shaped the ridge over millennia, and indirectly also tells what processes are irrelevant. Drifting of this sort is normal in any field of research.

1.2 Constructive Design Research

Our focus is on one specific type of design research, constructive design research. It is a phrase coined in 2011 by Koskinen and his colleagues (Koskinen 2011). While Koskinen et al. showed that constructive design research can be done in many ways, they did not, however, discuss its precedents outside design. The concept of Constructive Design Research emerged from having a set of research perspectives shed light on exemplars, concrete activities, and techniques of research performed across a number of design fields; some technically oriented, some arts and crafts oriented, and others action research oriented — all united by the interest of learning through constructing. The 2011 book pointed to principles behind constructive design research, and showed that it builds on three types of methodological backgrounds. Some researchers follow the sciences — mostly strictly those informed by experimental psychology — and conduct their work in laboratory style settings by de-contextualizing phenomena and studying it experimentally. Others prefer to go into the context and accept the fact that there is no way of controlling design when it enters real world. Yet others build on art and design and leave not only the scientific model, but also the social science model behind. Hence the main construct of the book, lab, field, showroom.

This is an ideal type, however, and only designed to capture methodological principles in design research. As soon as we turn our attention to the details of research practice, it begins to lose precision. In the preface of the book, the authors wrote:

> While talking about this book, many practitioners and researchers have found it immediately useful. One word of caution is required. Many people ask how their practice fits into the Lab, Field, and Showroom framework. However, we talk about practices that are seldom pure. In fact, Chapters 7, 8 and 9 look at how theory, research practice, and the social environment create commonalities between these approaches. These chapters have their origin in a 'deviant case': when we realized that it is impossible to classify Ianus Keller's PhD work under Lab or Field, we took a closer look at things that bridge researchers. (Koskinen et al. 2011. Preface)

This book steps into the field of constructive design research to describe one of its key characteristics — drifting — and how construction is related to research concerns of hypothesizing, experimenting, and evaluating. It goes to the minutia of practice. The main starting point that pushed us to this book was Stephan Wensveen's PhD thesis *A Tangibility Approach to Affective Interaction* (Wensveen 2005).

Wensveen built an affective alarm clock that tried to capture the mood of the sleeper to wake him up accordingly. His process started with cultural probes that inspired him; it went on to a process of sketching exploring a variety of design options; it continued with a construction phase of the alarm clock; and it ended up with testing the clock with users in a laboratory-like setting with sophisticated trigonometric methods (Fig. 1.2). Like our imaginary geologist's, Wensveen's approach drifted between methodologies depending on the current needs; instead of following his methodology strictly, he was flexible in his practice.

When we discussed his case, and started to look at other studies that defined the constructive methodology as we understand it, we saw that Wensveen is not an exception. Rather, his research is the norm in design research. Even hard-boiled field workers do usability studies with percentages, and design and HCI have seen more than enough attempts to turn cultural probes into scientific measurement instruments (to the chagrin of their inventors and proponents).

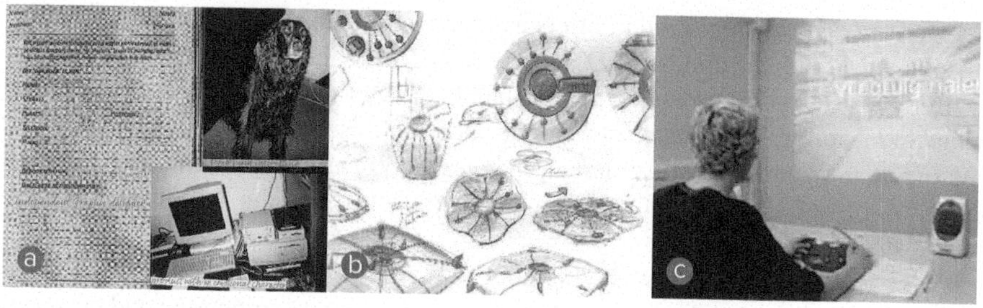

Fig. 1.2 Key activities in Wensveen's dissertation work (**a**) Cultural probes, (**b**) Sketching alternatives, (**c**) Laboratory testing

1.3 Drifting in Design and in Constructive Design Research

We are not the first persons who have paid attention to the fact that the practice of design almost inevitably pushes a project out of its original, planned envelope. It is a standard wisdom of the field of design that well-executed explorations will reveal that the problem that pushed a project on its way is not the true problem. When an architect gets a brief, he routinely writes a counter brief to show he has acknowledged the client's problem. When it is accepted, he usually gets more freedom to start the creative process. Similarly, industrial designers routinely negotiate a free option into their briefs even when the client wants to control the design process to an extreme.

This wisdom has found its way to design research too, which talks about changes in projects in several ways. In design anthropology, language usually builds on the idea of refocusing. The idea is that good ethnography has a tendency to show that the original problem given by the client in fact misspecifies it and can be dangerous if put to production (Kensing and Blomberg 1988; Blomberg et al. 1993). After Roberto Verganti's book (Verganti 2009), product designers in most design schools in the world taught students to aim at radical innovations, whether these be technological or innovations in meaning, which he equates with design. His case comes from Lombardy, and his ideas can be seen as an effort to put in words Italian design after the Memphis group. His 'radical innovation' also has commercial overtones which again show his interest in economics rather than design. In a recent book, Kees Dorst (2015) has followed Einstein and talked about reframing as a key design process. For him, if a problem has no solution in its original context, designers need to shift context and find a fresh way to reframe the problem.

The way in which we understand drifting in design departs in important ways from design world parlance, which we find too heroic to describe what happens in design when designers drift. If the standard for a successful project is that it produces something the world has never seen before, practically all projects are doomed to fail, and design gets pushed to an exceedingly risky track. The language or re-, or the language of radical innovation is to us too dramatic to capture the reality of drifting.

Drifting, as we understand it, offers a much more nuanced and sensitive perspective on what happens in design. Designers who drift by intention sometimes end up with radical ideas that are technical in nature. Sometimes these ideas are innovations in meaning. Sometimes they refocus and rephrase, and occasionally reframe. At times, they end up with minimal changes only. Design literature and folklore alike are full of examples of these kinds of processes, and the most heroic design world stories.

We know that what we are saying goes in many ways against the grain in some parts of design research literature, which usually builds on engineering and business discourses rather than on practices of design as performed in school of design. When discourses are borrowed from engineering and business studies, 'designing' is usually seen as 'creating innovation,' which by definition means bringing something new to the world. There is a virtual cottage industry around the subtypes of innovation, but usually they fall on the radical-incremental continuum.

The problem here, in our view, lies in the idea that the criteria of success is novelty; this is clearly not the case in many fields of design, where change is constant, but it cannot be always reduced to novelty. For example, craft is usually more about perfecting existing forms rather than producing things the world has not seen, and happens through redoing, upcycling, remaking, and repairing. The same can be said about many other fields of design, like graphic design in advertising or fashion, which usually works by reinterpreting what already exists. Secondly, equating design with innovation over-emphasizes design as an idea discipline and undermines the key competence and truly difficult part of the discipline: constructing a balanced and valuable product at the end. Thirdly, every working designer knows that most of the business is in tiny improvements that are sometimes almost cosmetic. Changing colors in an action figure or updating a heat exchanger to accommodate new fittings is hardly a radical innovation, but it does create something new — and maybe not least changes the practices around it.

Reducing design into a field defined by one variable like innovation, we believe, is clearly a too radical way to understand design. This is true of constructive design research alike. The history of constructive design research has seen few radical innovations and reframings. Instead, research mostly proceeds by producing small changes in our understanding of design. These changes are sometimes methodic, sometimes conceptual, and occasionally even theoretical. Often, they are more like brokering: for example, taking existing methods from user-centered design into studies of autonomous vehicles.

The notion of drifting opens the door for seeing many aspects of constructive design research that existing frameworks miss: those small-scale changes professional designers do in their studios and design researchers do in their research institutes, and those molecular changes Andrea Branzi (Branzi 2013) has been talking about. We claim that the notion of drifting opens doors to inquiries into many important yet overlooked aspects of design.

In our use, the word drifting comes from a conference paper by the Swedish designer and philosopher Johan Redström (Redström 2011), who looks at the relations between experiment and program in constructive design research. In his view, in particular the early experiment, both causes drift in the articulation of the program, and serves to stabilize the program under which a knowledge contribution is defined. We think there is more to the act and notion of drifting than what he points to, however. Although our study shares part of its genealogy with his work, our argument differs from Redström in two ways. First, his unit is larger than ours. In his parlance, research programs drift. We believe every design research project drifts, and if it does not, it becomes what design researchers usually call clinical research, application of knowledge to individual cases, following the American design scholar Richard Buchanan's (2001) classification of research into clinical, applied and basic research.

Another difference is that our notion of drifting is intentional rather than structural. We do not see design researchers as driftwood being pushed around by the winds and the currents when it searches its way towards the sea. Instead, designers drift intentionally, and do this in many ways. They negotiate slack into their contracts. They make counter briefs. They sketch dozens of ideas to explore conceptions and possible futures and keep them underspecified and sketchy to invite for co-production and avoid untimely criticism from clients and potential user. They create models to mislead customers temporarily to make sure they can work on a better design in stealth mode until it is finished (Esko Kurvinen 2007). Drift is thus not only something that happens until a certain research program has been stabilized, but something that is fundamental in research.

1.4 New Experimentalism as a Lens

A major inspiration that led to this book is the notion of new experimentalism in philosophy of science. It was made popular by the philosopher of science Deborah Mayo. In her book *Error and the Growth of Experimental Knowledge* (1996), she argued that scientific experiments as such create new knowledge. Her concept led us to ask how design experiments create knowledge. The notion has been implicit in design research (Redström 2011; Bang and Eriksen 2014; Brandt and Binder 2007), and it is certainly acknowledged in abstract, but it has received little detailed attention so far. Yet, it raises important questions about the role of design work as a site of creating knowledge.

For us Mayo's new experimentalism in constructive design research consist of three main activities: hypothesis construction, experiment(s), and evaluation — a structure that is also reflected in the mid-section of this book. The process is most often iterative, especially at PhD level. Of course, every designer experiments, and experimentation characterizes any design process from its very beginning. Designers sketch ideas, combine them, build mock-ups and when constructing prototypes, typically create several models before deciding which one works best — experiments as trails of potential futures.

Experimentation of this sort is closer to professional design practice than experimental research in science. It does not have a strict methodology, and even if it has, this methodology is not abstract, not grounded in stochastic processes, nor is it methodic in the sense that there are right and wrong ways of doing experiments. Instead, if we are right, constructive design research can be seen as a form of experimentation with forms, materials, spaces, mechanisms, and programs. The only way to know these things, crucial to design, is to try and err. In line with the promotion of the theory unbound experiment promoted in Mayo's new experimentalism many of these experiments come from the needs and interests of design rather than theory.

It is good to note that experimentation may take several forms. Following the taxonomy of the 2011 Koskinen et al. book it can follow a laboratory-style format in which variations of the same design are treated as hypotheses, and evaluation studies determine which one offers the best solution. When done in this way, research tends to use statistical techniques and it tries to achieve a degree of control over the circumstances in which the study is done. At the other end, critical designers create objects that become props in conversational encounters with the public, and there is no attempt to gain any degree of control over who is present in these encounters, and who gets to talk.

Seen as new experimentalism, constructive design research has a few characteristics. It is grounded in research discourse and contributes to it, but it is also playful. Like any design, it can be inspired by almost anything, but when it leaves inspiration, it is informed by what other researchers have done. The contributions that are derived from experimentation are filtered by four traditions of knowledge we shall describe in this book.

1.5 Explicating Drifting

This book builds on a corpus of PhD level design research collected over the years systematically from Northern Europe for further information. The corpus is described in Chap. 10. All the work builds on a constructive approach, and shows a variety of ways in which drifting can happen. Most of the work comes from the borderline of product design and interaction design, it often has a base in HCI, and sometimes it ventures to textile design and craft, which has followed the lead of product design and taken a research turn especially in North-Western Europe. This corpus of work is too large to be studied systematically, but when writing this book, our corpus consisted of 67 well-referenced and influential theses. They come from a variety of methodological and theoretical backgrounds, and the authors have supervised several of them.

Our method of analyzing drifting has been classic explication. The method consists of examining a question against the corpus, and then building an analytic framework that describes observed variations in the corpus. Depending on the question, relevant sections of the corpus vary; for example, when analyzing experimentation, we focused on the section of design experiments; while when studying evaluation, we mostly focused on the final section of empirical work.

Why are we selecting PhD level work for analysis? The reason is that design research is coming of an age in this work. Almost every writer in our corpus has a design background, and every single book describes at least one design experiment and uses it as a vehicle for creating knowledge. In doing a PhD, design researchers have encountered methodological and theoretical questions coming from more established disciplines, they have found ways of reconciling these with their own beliefs, and they have had to describe these ways in detail. Our corpus tells about these struggles, and also shows ways out of them.

Thus, even though it is not a highly polished body of work (PhD theses also tell about supervisors, departments and local research cultures), it is a rich place to look for if we want to make sense of drifting in constructive design research.

Our method is mainly typological in the same sense as architects talk about building in terms of typologies. They are ideal types that capture central tendencies but many cases do not fit them perfectly.

The method has its basis in work leading to this book. In 2015, the first author published a paper that classified the ways of drifting into five main categories. One of these was 'expansive' in which 'there are no strict successive or linear orders or directions to follow. Experiments and learning from this will contribute new knowledge, as the area is explored' (Krogh et al. 2015: Sect. 3.4). The main example in this paper was Christian Dindler's thesis exploring how through Participatory Design one may develop 'Engaging interactive environments':

> Rather than deepening our knowledge of a domain, this method widens our perspective and extends the concerns we, as designers, should include in our praxis… Dindler expands our idea of what 'engagement' might mean in interaction design through three diverse experiments including designing engaging artifacts and developing methods for participatory design centered on engagement. (Krogh et al. 2015, ibid.)

One good way to understand this method is an image of a biologist who explores a meadow on a yet uncharted island — the subject of interest is very well delimited. He can explore it in several ways, of course, but one good way is to explore it systematically by starting from one edge and then expanding the analysis to other areas, one chunk at the time, until the whole meadow has been covered (given the budget, of course).

The main limitation of the characterization of the ways of drifting is that they are mostly based on an analysis of research carried out in a North-Western Europe covering design practices from product design to fashion and to interaction design. While we are aware of these limitations, we also believe explicating models is a fruitful way to develop disciplines. It has sensitized us to the diversity of research approaches and by doing that helps us to avoid some of the Siren songs that tell designers simplistic versions of how research must be scientific. The models we identify and describe in this book are probably not exhaustive, but this is not the purpose of the book. It is not an encyclopedia and it is not a handbook. It is the result of our study of the corpus described above, following the spirit of the statistician George Box, who noted that 'all models are wrong, but some are useful' (Box 1979, see also Box 1976). By leaving things out, models gain clarity and expressive power.

1.6 On Terminology and the Structure of the Book

The main claim of the book can now be gathered. We believe design can produce knowledge, and we know this can happen in several ways. We believe, however, that existing literature has neglected an important, though typical design research activity we call drifting. In producing knowledge, design researchers drift from one idea to another and from one methodology to another, and they do this intentionally. Most accounts of methodology, in our opinion, are too linear and logical to acknowledge this activity.

In this book, we use several terms that have several uses in design. Mostly our use should be clear in context, but two words require some explanation. When we talk about 'craft', we usually mean skilled practice, but also refer to craft research, which has become a serious field over the last 10–15 years mostly in the UK, Scandinavia, and also in Australia. When we talk about 'dialectic tradition,' we mean a host of design practices in which designers work with people and let their viewpoints lead drifting. Participatory tradition in our parlance covers several approaches, including participatory design (Pelle Ehn 1988), empathic design (Mattelmäki et al. 2013), several forms of co-design (Mattelmäki and Froujke Sleeswijk Visser 2011), most versions of co-creation (Sanders and Stappers 2008), and various types of 'dialogical' design (for instance, Tuuli Mattelmäki 2006; Ezio Manzini 2015). These have different origins and although participatory design has been historically important in bringing users into the design process, it is by no means the only approach that has done that.

Design is an ambiguous term, and there are no approved standards to naming the products, processes, communities, discipline, approaches etc. of the field. Here is a brief declaration terms used in the book. When we speak of *skills and practice of design* we mean the acts deployed to make a process move. There are different traditions for terming products of design, some will use 'design object' and others 'design artefact.' We attempt to use *artefact*. However, we use '*design object*' if other researchers have used it. The term *object of design* is the concrete instantiation of an exemplar e.g. a chair which is currently *subject to design*. The evidence of explorations in processes of design we term *experiments*.

The book is divided in three parts collectively showing and arguing that drifting is a characteristic of constructive design research:

- *Chapters 1, 2 and 3* ground our discussion of drifting in the concept of knowledge and answer why we think the concept of drifting is valuable. In addition to this introductory chapter, Chap. 2 maps some of the changes that have taken place when design practices have been repurposed into research tools, while Chap. 3 maps how concepts of knowledge have played a role in this transition.
- *Chapters 4, 5 and 6* are the heart of the book and focuses on epistemology of the constructive design research. Its question is how drifting happens. It builds on a PhD program one of the authors has been building in Jutland, the Western part of Denmark. This program trains PhD students to frame their work through three concepts: hypothesis, experimentation, and evaluation. Chapter 4 elaborates on

how in the early parts of any design process things are in the air, or in research world language hypothetical. Chapter 5 described how when some ideas have been selected for further development, they are being turned into design objects that are seen as experiments rather than prototypes, which is the typical other alternative in design research. The word experiment carries a connotation we like: it frames design research in research language, and shows that ultimately design objects are only experiments, not more than that. The final part of the program consists of evaluation, which is the topic of Chap. 6. Evaluation comes in design research under several names. In the early days of constructive design research, most researchers talked about evaluation or (field) testing, but more recently the language has taken a military overtone, speaking of deployments instead.

- *Chapters 7, 8 and 9* zoom out from the details of research and looks at emerging ways of doing research. The question these three chapters pose is: where is constructive design research heading towards, and who beyond design research might benefit from it? By analyzing some of the trends around it, we use this part of the book to explore how the constructive practice is likely to develop over the next few years. These chapters build on our experience in design research. For years, constructive design research used to be in the sidelines of design research, which was dominated first by art history and later by user-centered design. By now, it has become mainstream. In the process or becoming mainstream, it has also shaped other design research practices. For example, some practices have become more constructive. This tends to be the case in user-centered design, which used to be mostly about the early stages of the design process (in our language, about hypothesis), but now routinely covers the experimentation and evaluation. However, these changes are beyond this book.

Chapter 2
Shift to Knowledge

In the previous chapter we outlined the background of this book — constructive design research — and unpacked how constructive design research, being partly based on design practices, needs to pay respect to both professional and academic worlds of design. We introduced the concept of 'drifting' as an ostentatious way of articulating how processes of design change and adopt as new knowledge is build and the need for reframing its concerns emerges. Stated metaphorically: not as driftwood, but as in car rally; intentionally and controlled. The chapter also outlined how constructive design research in increased self-efficacy bridges ideas and ways of working from other disciplines in its particular ways.

This chapter will point to the change in the character of knowledge build by and for design when its base shifts from practice to research and studies drifting in the context of these arguments.

Historically fields like design and architecture have built knowledge by exemplars. In a knowledge base of this kind it can be hard to be specific in where a foundation of choice is made by tradition or by substantial new knowledge. Furthermore, historically, the language of design, architecture and crafts has focused on artifacts and it has located change in the personality of the designer and in his artistic explorations. In contrast, the peasant in ancient time had the honed and respected capacity of discussing the quality of a wheel with both the blacksmith and the carpenter producing the wheel. Such respectful relationship between a crafted producer and a sophisticated user today is found in arenas of passionate use, like in musical instruments, yachting, fishing and others where experience is the main base of use. Passionate musicians and yachters are capable of fixing most broken things on the fly and are willing to travel hundreds of miles to get the right parts and service for their equipment. Adding or at times even substituting this form of communication with written texts would profoundly shift the base of design from experience to knowledge. When this shift happens, notions of what is good and proper come to be construed as questions of knowledge: validity, reliability, evidence, correctness, truth likeness, and even truth, experience gets pushed to the passenger seat.

P. G. Krogh, I. Koskinen, *Drifting by Intention*, Design Research Foundations,
https://doi.org/10.1007/978-3-030-37896-7_2

The implications of this shift to knowledge and knowledge communicated dominantly in text in design research has profound implications. Usual measures of knowledge, like the Platonic idea which defines it as true, well-justified opinion, are ideals that help to understand design only in part. The same goes to scientific ideals too, we believe. These can be realist, for example, building about notions of evidence. They can also be consensual and build around peer judgment; instrumental building around what works; or critical building around emancipation from prejudices and power structures (Habermas 2015).

As this chapter will show, however, the shift to knowledge has happened in a number of ways. Design researchers interpret knowledge in many ways, and these interpretations at least partly define the ways in which they approach research. They build on several philosophical and pragmatic positions. At the other extreme, there are researchers who have contributed to knowledge while phrasing their creative process in terms of their unique personality, or who rely on ambiguity as a critical resource that helps to create knowledge.

2.1 Knowledge as Method and as Aesthetic

At the moment, constructive design research most dominantly takes place in three communities of research: engineering; design that builds on arts and crafts; and human-computer interaction. Rather than representing three distinct communities, they sit on a spectrum with several overlaps. In the following we illustrate this by pointing to exemplars from two extremes of the spectrum in terms of how much importance they place upon methodological rigor and transparency.

The engineering communities usually locate knowledge in correct application of research methods. This approach has a long history. In design, the best-known examples are the design methods movement exemplified by John Christopher Jones's *Design Methods* (1992), which described and classified methods by their fit to a design process. For Jones, design methods could be described by a matrix in which methods changes as the process went on from early stage studies through data gathering and innovation all the way to evaluation. A more recent example is the work by Lucienne Blessing and Amaresh Chakrabarti (2009). Their approach is refined, but ultimately wants to offer a unified language over the design process. Their model describes design as a series of three studies. The first is a descriptive study that consists of methods like desk research and observation; the second is a prescriptive study that generates a concept and a design; the third study is another descriptive study in which this design is evaluated. In the second study, the method covers industrial methods like the waterfall methods in software development rather than research methods only. The methods in the first and in the third phases are familiar to any scientist and social scientist.

Blessing's and Chakrabarti's sophisticated model illustrates well how design can be understood as a rational, well-controlled process. In it, drifting can happen in the very early phase of a project, but the rest of the process is kept focused to partitioning the problem into tasks and subtasks that have a solution and build up towards a well-functioning design. Drifting late in the process becomes a failure in planning.

This model has undeniable merits. When methods are clear and everyone follows them, it is easy to inspect a piece of work to see whether it is correct or not. And, if everything is correct but something still goes against expectations, researchers know they are on something. By basing knowledge claims on methods, research community forms an 'intersubjective'[1] community that is able to locate its problems quick and produce results that can be defended even against demagogues who contest the validity of scientific knowledge. Well-cited rationalistic models of design abound in engineering and in information systems (see Hevner et al. 2004).

Despite its merits, this tradition also has its share of pitfalls. It is specifically contested by Gaver (2001), who distinguished between epistemological accountability and aesthetic accountability (a notion we will return to in Chap. 3). This distinction was designed to show that design can build on practically anything and have no methods, and still be valid. The following quote is from *Presence Project* (2001), which does not have clear first author, but the writing is Gaver's:

> Science is epistemologically accountable. In pursuing its goal of producing causal explanations of the world, what distinguishes science from similar endeavors (e.g. religion) is that the methods employed to establish facts, construct theories, and relate the two are used to justify the validity of one's work, not the results of these methods. No matter how plausible or desirable a statement or theory, it does not count as science unless the basis for asserting it is clear and veritable. Conversely, no matter how implausible or even dangerous a statement ('the earth revolves around the sun'), it is a scientific fact if the means used to assert it are sound. In principle, epistemological accountability allows science to establish causal models that are counter to prevailing opinion thus defending progress against prejudice. But epistemological accountability also implies that work can be scientifically valid despite obvious or trivial[…].
>
> There are several implications of science's foundation on epistemological accountability. First, scientific methods must be articulate and precise. These methods — controlled empirical investigations, statistical analyses, logic and deduction — allow the chains of inference used to posit facts or theories to be examined and verified by independent researchers. Second, facts are, by definition, objective and replicable. That is, they do not depend on any given person's perception or beliefs, and will be found in a given set of circumstances by anybody who does the looking and whenever they look. (*Presence Project* 2001: 202–203)

At least two types of historical evidence suggest that Gaver is correct. On one hand, the fate of what is known as design methods movement tells he hits the target. The movement tried to put design on a methodic foundation in the sixties, but it was discarded in about 10 years as a failure by the founders of the movement (see Tomas Maldonado 1984; J.C.Jones 1991; Christopher Alexander 1971). Also, the history of design speaks for Gaver. More often than not, designers who have shaped the discipline have had their background in theater and staging (Henry Dreyfuss and

[1]We use the term of 'intersubjectivity' here to point to the interactional agreement happening among stakeholders in a community, allowing it in research discourse to point out and defend particular ways of producing and defending knowledge.

Normal Bel Geddes), sculpture (Marcelo Nizzoli), craft and carpentry (Hans J. Wegner and Tapio Wirkkala), and they have been reading non-scientific philosophy like theosophy (Norman Bel Geddes).

Having said this, the methods argument is obviously important for professional practice, which needs standardized methods. It is also important to industry in which a product development process may involve thousands of people. There has to be a way to make sure everyone in this process can know what others are doing. To coordinate the work of hundreds or thousands of people, standard methods and processes are a must. The need for a process is a matter of control rather than knowledge, however, and moves discussion to from research-world concepts to industrial relations.

In the following we look at the position and origin of knowledge as it is built in constructive design research.

2.2 Artifacts as Knowledge

As the quote from *Presence Project* shows, there are influential voices in design research community speaking for the idea that design knowledge resides within an object. It has become particularly popular in communities that focus on research-through-design (Christopher Frayling 1993) and human-computer interaction. The first paper that raised the question of how design artifacts can become a part of knowledge production was published in 1989. It was written by John Carroll and Wendy Kellogg, and it claimed that artifacts are 'theory-nexuses.' A dictionary definition of nexus is a series of connections linking two or more things. In this sense, artifacts like a computer mouse are nexuses: they link several ideas and theories together. They do this seamlessly in the sense that once an artifact exists, and it can be broken down into its constituent components, these components are no more its meaning. The meaning is the artifact and what it can be used for, not what went into it.

One implication of their argument was particularly relevant for design. Carroll and Kellogg observed — perhaps somewhat regretfully — that what they say means that theory and practice may sometimes swap roles. Practice takes the driver's seat. If a researcher wants to study why a computer mouse works, he need to have a mouse first. It cannot be studied as an imaginary construct because of its nature as a theory-nexus. As they write:

> Indeed, interface innovations frequently lead HCI research rather than following from it in the conventionally assumed flow of 'technology transfer' from theory to application. Often, research returns again and again to the interpretation of particular artefacts and techniques. The evolution of the concept of 'direct manipulation' as a psychological analysis of a collection of HCI artefacts provides an excellent example of this… The theoretical work substantially postdates the implementation of the concept in running systems. (Carroll and Kellogg 1989: 7).

In design and HCI research communities, this argument gave credence to the idea that knowledge may reside in an artefact and cannot be reduced to those theories that supposedly produced it. One consequence of this was that theory got de-centered, which paved the way for seeing design practice as the main site of knowledge production.

It took several years before this was understood and accepted widely. An important step took place in London after Christopher Frayling became the head of the Royal College of Art. Before him, RCA's previous head Jocelyn Stevens had disbanded the school's Department of Design Research after Bruce Archer's retirement in 1988 (McIntyre 1995). At that point, Stevens saw research as irrelevant to design: it was too theoretical and too focused on explaining how designers think rather than on the products and merits of design practice itself.

In his attempt to rebuild research at RCA, Frayling looked around his school and wrote a short paper at the end of which he put some pieces of work, which he consider research, under three categories. One of them was 'research-through-design,' by which he meant the work designers do in their studios while aiming to produce knowledge rather than products (Frayling 1993).

Written in a small working paper and never meant to be taken very seriously, this paper was largely forgotten until some of the work Frayling mentioned matured into contributions at PhD level. Most notable was Anthony Dunne's PhD thesis, published in 1999. Dunne was designer by training and his thesis was an exploration of how to find imaginative ways to design electronics. His approach became particular evident in his later book *Design Noir—The Secret Life of Electronic Objects*, which he co-authored with Fiona Raby. In this book, they designed objects to serve as of 'placebos,' furniture-like objects that probed those imaginative everyday tactics in which people cope with — to some the troubled experience of —electromagnetic radiation (EMR). The project playfully explored the ways in which beliefs about sensitivity to EMR. The study was interested in 'peoples' attitude and experiences of electromagnetic fields in the home,' and as Dunne and Raby write, 'the placebo objects [were] designed to elicit stories about the secret life of electronic objects — both factual and imagined' (Dunne and Raby 2001). *Design Noir* was not interested in declaring or reflecting upon the process of conceiving and producing the objects. The focus was on documenting the potential role such objects might obtain through use in particular people's daily life.

Design Noir is a good example of how Frayling's concept opened a new way of doing research at RCA. Although the concept of 'research-through-design' was (and is) ambiguous, it opened the doors to the idea that artefacts of design can be a form of knowledge of its own, and gave this idea legitimacy form the highest levels of an important art school. After his essay, it was easier to argue that some aspects of reality can best be approached by doing design rather than by studying designers, as in Bruce Archer's late research. While his terminology can be criticized, his concept had the merit of opening the door to the argument that design explorations may be a form of knowledge of its own. Following him, it was possible to argue that the site from which design produces knowledge was the act of designing, not cognitive processes in design thinking and problem solving. This view went against the mainstream of the time, which sought to find general decision-making patterns, but it rose to prominence in Dunne's generation.

The view caught fire and went through a series of transformations over the next few years. Some researchers continued to defend the idea that design artefacts are things of their own and should not be seen as reflections of theories and only open to science. Most notably, the closing credo of Gaver in *Presence Project* (2001), referenced earlier in this chapter, argued that design is accountable aesthetically rather than epistemologically and an artefact can be valid even if it does not build on scientific research. This view has a clear affinity to the main credo of *Design Noir*, in which Dunne and Fiona Raby explicitly told that their study is not scientific in nature.

It is also along these lines that (Koskinen et al. 2011), identified the approach of 'Showroom', in which knowledge becomes a series of short-lived conversations about issues designers want to raise, and possible consequences of these conversations. There is no urge to verify whether these conversations lead anywhere, nor an urge to study how the long-term effects take place. In fact, the very idea of such research is treated with skepticism and research is usually designed to question existing beliefs. The main criterion of success is whether research opens doors to new ideas and concepts.

These perspectives are 'end-product-oriented' in that they put emphasis on the design outcome. With regard to the corpus that serve as foundation of this book this approach is found e.g. in the works of Niedderer (2004) with her interest in 'Performative Objects' (e.g. Social Cups, the capacity of design objects to invite for particular actions), and Kart Summatavet (2005) who explored jewelry in three stages of a woman's life (the childhood, maturity, and old age) and exhibited her work in museums in Helsinki and Tallinn. In terms of research venues, it has found support in craft research, which has several roots, including Michael Biggs's interpretation of Ludwig Wittgenstein's philosophy and Frayling, but also partly on a variety of other sources (see Biggs 2000; Rust, Mottram and Till 2007).[2] Its most significant communities are probably *Craft Research, The Art of Research Conference*, which has been running since 2007 (see Mäkelä and Routarinne 2006), and *Research through Design Conference*, a brainchild of Gilbert Cockton, who organized the first conference at Northumbria University in 2012 along with now defunct *Designing Pleasurable Products and Interfaces*. These venues have sought ways to put products into the very centers of design research. Notably, all these venues have their beginning design schools like the former Chelsea College of Art and Design in London, the former University of Art and Design Helsinki, and Northumbria University's design school.

[2] Rust, Chris, Judith Mottram and Jeremy Till 2007. *Practice-led Research in Art, Design and Architecture*. AHRC Research Review. Available at: http://arts.brighton.ac.uk/__data/assets/pdf_file/0018/43065/Practice-Led_Review_Nov07.pdf, Accessed 9 July 2017.

2.3 Frameworks and Programs: Discourse as Knowledge

Another interpretation of Frayling's short paper was daylight in human-computer interaction. The main characters here were in Pittsburgh, where Jodi Forlizzi, John Zimmerman, and some of their colleagues created a novel interpretations of Carroll and Kellogg's and Frayling's texts (Zimmerman et al. 2007; Zimmerman et al. 2010). They appropriated Frayling's concept and like him, argued that design can be a source of knowledge, but posed the question on procedural grounds rather than located knowledge in objects. Their answer built around beliefs in a design-oriented HCI environment in which constructing devices was an accepted way of proving an idea, but building things as such was seen as design or engineering only. In contrast to the London group's belief in the artefact, they situated knowledge in discourse around the artifact. They tell designers to build frameworks around their work, and introduce these frameworks into discourse in which knowledge is being debated, criticized, and ultimately stabilized.

This view seems to dedicate little attention to actual design work, but this is of course a misinterpretation: design practice and producing resulting artefacts is a necessary, though not sufficient step in producing the frameworks discourse needs. This view clearly put design first, and sees frameworks as an outgrowth rather than the final point of design research (though it remains unclear whether in the long run this view, if accepted, would turn design into a science). Ultimately, the view developed in Pittsburgh encourages design researchers to construct designs and to build frameworks out of these designs to tell to others how they should be understood. This interpretation of Frayling places the locus of knowledge creation into discourse in contrast to the artifact.

One benefit of this view is that it provides a way to manage something the philosopher Don Ihde once called The Designer Fallacy (2006). By this term, Ihde pointed out that designers sometimes have a bad habit of thinking that people who see or use their creations understand them the same way designers see them. This, however, seldom happens: artefacts are used in many ways, and often many of these appropriations fall outside the scope of the designers' intention. No doubt, there is also a Design Researcher Fallacy. If objects are left on their own, they can be interpreted in many ways, and it is highly likely that many of these interpretations are different from researchers' intentions. If constructive design researchers propose a framework at the end of their work, it provides a base for interpretation and shows how they prefer their work to be seen. Such framings facilitate preferred interpretation of sorts, and provide a place to start critique.

This vision has been taken further in yet another line of interpretation. Working from interaction design rather than HCI, Thomas Binder and Johan Redström (2006) introduced the notion of 'programs' into design literature. They argued that to understand progress in design, attention should go to programs rather than individual projects or even frameworks. They used the term program in architectural sense, however. An architect studies a plot, analyses it, and creates a program for building it. Slightly later, they worked with Koskinen, who defined program in terms of Imre Lakatos's philosophy of science, in which the locus of progress is located in theoretical debate. This line of work has been extended through the book *Making Design Theory* by Redström (2017).

If they are correct, knowledge, in this view, may be created in individual projects and in discourse around them, but even more importantly, it is being created over a series of studies. In their world, one study leads to another, which works on problems discovered but not solved in the previous study, and it simultaneously prepares the next one. Their work was meant to pay attention to how design creates knowledge at this abstract level. Just like in Forlizzi and Zimmerman's interpretation, the artefact is a necessary component of knowledge creation, but second to the discourse that it is meant to generate.

One implication of this argument is that some studies may in fact be wrong, but they may still contribute to the growth of knowledge if they contribute to the bigger picture that is developing. Also, no study is decisive on its own: we know something because of the cumulative weight of several studies, as in research on nutrition and the health effects of tobacco. In the latter, for instance, there is no doubt about the fact that tobacco is harmful for humans. Yet, it is impossible to take one human being and tell that tobacco was the cause of the cancer that killed him. Still, given the combined weight of thousands of studies, the skeptical argument that questions the link between smoking and several diseases has been ruled out.

Although the scale of research is considerably smaller in design research communities than in medicine, it may well be worth taking the research program argument seriously. We can see, for instance, that James J. Gibson's (1979) ecological psychology is a fruitful theory to guide practice of design not because of any single study, but because taking his psychology seriously has led to interesting results time after time across disciplines, communities, and domains of design (product design, information science, interaction design across lab and field based studies, across home, work and leisure studies).

The way in which debate is understood takes several forms, however. IT wan be taken as an accumulation of facts. A research group particularly recognized for its work on Gibson, the *Designing the Quality of Interaction Group* in Eindhoven led initially by Kees Overbeeke, represented in our corpus by authors like Joep Frens (2006) and Stephan Wensveen (2005). For them, knowledge depends primarily on methodology. Their aim is to build universal knowledge that would gain its undeniable character from methodic precision. Those coming from the empathic backgrounds, however, tend to see knowledge in terms of growth spurts much like anthropologists (Geertz 1973).

2.4 Knowledge as Participation: How Users Motivate Drifting

Whereas the two prior sections deal with origins and positions of knowledge production in constructive design research where researchers are in control of drift, the inclusion of users changes the game. When users, stakeholders and passers-by are included in a co-design process, they introduce a whole different set of non-controllable sources of that drive drifting.

Involving users in design processes rely on different practices ranging from human factors inspired approach where recording (physical or cognitive) conditions serve as requirement for design such e.g. Paul (1954), to a wide array of techniques, methods and approaches for involving laypeople as co-designers, reported well by Sanders and Stappers (2008). Looking to the early days of cooperative design — known in English as participatory design — of information systems, the engagement of users pointed both to the design of systems and prospectively to the education of workers in the co-designing future systems (Bødker et al. 1987; Bødker and Kyng 2018). Design processes became in part an activist intervention and in part requirement specification for designing information systems. Looking at our corpus, we find examples like Christian Dindler, who represents this tradition.

Drifting is driven by what designers learn from people. The way in which these drifts happen depends on the design approach, however, a good way to illustrate this is the difference between empathic and participatory design. Both are Nordic geographically, but although they are sometimes lumped together, they have significant differences.

Empathic design tells designers to study users and their experience in context, and define whether they want to be neutral observers (like a design anthropologists might), or participate in the experience (as a designer might). What produces valid knowledge for design decisions does not reside in methods or designers' imagination (whether fictional or theoretical), but in user experience distilled through the designer's understanding of it. A softer version of this argument is reminiscent of how Dreyfuss and Tilley (2002) saw ergonomics: as a source of inspiration and as a checklist rather than as a source of knowledge to be followed because it speaks with scientific authority. In this line, research may reframe the original question, but it becomes a starting point for the design proposals. Learning from user studies should not dictate design, but directs it from within the design process. This is in line with how most practitioners see research: as an important source of insight, inspiration and knowledge, but also something to be wary of. People do not always know best, and user studies do not answer all questions designers have.

Following the participatory design paradigm, we see yet another approach to users. In the most political approaches to participatory design, the key objective of a design activity is mutual learning between professional designers and amateurs (the stakeholders and users). Design has an activist component to it, and it attempts not only to understand but also to educate the amateur. The purpose is to help the amateur to take command of (technological) options, and to exercise his or her democratic right to influence the design of objects, technologies, services, environments and so on, so that their quality of life improves. Participatory design is rooted in co-operative design in the Scandinavian countries in the mid-seventies. Among others, it was inspired by Jane Jacobs's attempt to democratize urban planning in her *The Death and Life of Great American Cities* (Rothman 2017), the Brazilian educator Paulo Freire, Karl Marx, and the Norwegian computer scientist Kristen Nygaard. Following Nygaard and others, people like Pelle Ehn, Morten Kyng

(1979) and Susanne Bødker (Bødker et al. 1987) educated workers like typographers whose job was rapidly changing and widely substituted by the advent of digital technologies by developing technologies that took their interests better in to account. Lately, Bødker and Kyng (2018) have been reflecting on how participatory design is still a very relevant approach and appropriate response to the big issues arising from the ever increasing digitization of society. They are making a call to researchers to be activists in the pursuit of lasting democratic visions on technology use and inventions. In the perspective of drifting, the interesting aspect of the call is that the activist designers ended up spending much more time on educating the user to take active participation in the design of future work places than designing systems. When looking at this, processes of design have drifted from the pursuit of concrete results related to inventions, to become a process of learning aimed to educate society and qualify democratic influence.

A recent example is the perspective of 'patient empowerment,' which aims at spurring patient resources in healthcare by involving patients in it. The approach, though modernized, is very much in line with the thoughts and ideals of the early co-operative design in Scandinavia (Bødker and Kyng 2018). Artefacts resulting from this process become not only solutions to challenges at hand, but also vehicles for learning. There are no predetermined criteria that a solution need to live up to, though. The quality of the process is its capacity to drift and continuously adopt to the needs, skills and competences of the participants. The intentional drift is directed to maximize the building of knowledge among all participants.

Seen from the standpoint of drifting, participatory approaches like empathic design and participatory design have pushed aside the scientific ideal that method provides incontestable knowledge. Rather, each one of them gives priority to interaction with people, although in different ways. Some views claim that drifting must happen in the early phases of design research (see Redström 2011). According to this view, once researchers have found something, these findings become the base of the design process. This view as such is perfectly logical: why waste money on research if designers have full freedom as soon as they start their work? Other views have more tolerance towards drifting in the early phases of development projects. They tend to see the early phase and its outcomes in terms of inspiration, treat the findings as a priming process, or even as a sensitizing process, and tolerate even large drifts if the design process so tells. This view is well in line with a pragmatic interpretation of design (Peter Dalsgaard 2014), which avoids categorizing things into classes that could be turned into recipe-like methods and processes. Yet, neither view throws doubt on the central claim that sees user studies and involvement as a main driver behind drifting, and in some cases as even the main driver.

Both perspectives can be seen as adaptations of user-centered design methods — after all none of the above is concerned with, say, sustainability and the protection of nature reserves. These adaptations have pushed design away from its historical base in which design objects are seen as artful inventions of the designer shaped by his or her personality. They have also pushed aside the idea that design practice needs to become a science that builds its authority on methods. Current views on how research shapes design are well in line with the contemporary standards in qualitative social science and in the humanities, where methods are more malleable than in the sciences, and the main focus is on debate than on theory.

2.5 What to Expect From Constructive Design Research: Re-Articulation

By now, there are several approaches to constructive design research. As this chapter suggests, we see it as a term that can function as an umbrella that covers at least four different ways in which design researchers understand knowledge. We wanted to touch the concept of knowledge because it is often left implicit in design research, even though the way in which it is understood sets the ethos of research, and this ethos can vary from scientific to humanistic and even artistic. Needless to say, our focus has been limited to our topic, constructive design research.

After over 20 years of attention paid to how practices of design can generate knowledge, we have learned to see it as a multi-faceted phenomenon that happens in several ways. Our survey shows that although constructive design research produces artifacts, it usually detaches these from the designer's person (or company) and in this crucial respect differs from the professional practice (there are other differences too of course). The site of knowledge production is sometimes in methods, sometimes in the artifact, and sometimes in the theoretical, experiential or activist discourses generated around these artifacts.

This is a confusing way to paraphrase what design research can do and for designers, it opens the door to bad design disguised as research. For researchers, it opens the door of bad research disguised as design. But, at best, both sides are happy. This dilemma waits anyone who ventures into constructive design research.

The main line is clear and common. We know that the core is an artifact. We also know it can be treated in different ways methodologically. We know, furthermore, that different researchers and groups have developed several ways of telling how the artifact leads to knowledge: some stress it as such, some see it as a starting point in discourse, and some find their home between these extremes. Most constructive design researchers end their papers with design implications, design guidelines, sensitivities or, as we have noted in the previous section, frameworks. Design research has also seen a virtual research technique industry, which has produced methods, frameworks, meta-level discussions about these frameworks, and methodological explorations. Underlying these discussions is a shared belief that design research leads to better ways of doing design if it builds on design practices instead of exclusively borrowing its research practices from other fields of inquiry.

Where do these differences lead us as a community? This is unclear, but there is at least one clear vision in literature. Kees Overbeeke and his colleagues dreamt of using their craft as a stepping stone for building mathematical models that would over time become the intellectual core of design. In his thinking, design artifacts became test benches of theories, and their purpose was to serve as physical hypotheses. In the following quote, Overbeeke and his colleagues talk about interaction design, but the thought can be generalized to all design. If they were correct in the sense that design would be able to create mathematical models, design would become like any science. Ultimately, research would not be needed: the equations at the core of the discipline would do.

> Design research resembles research in, e.g., psychology in that it has a minimum of controls built in when exploring the solution when testing variations of solutions. Therefore we wrote above: 'we have kept the devices simple, pure and with resembling aesthetic appearance.' This makes it possible, to a certain degree, to isolate and even manipulate systematically critical variables. What is of the utmost importance, we think, when testing the variations, is that the user's actions on the prototype are recorded… In the future, we will use these data to construct mathematical models of actual interaction. (Overbeeke et al. 2006: p. 64-65])

The vision of Overbeeke and his colleagues is bold but it has company. In a recent article, Stappers and his colleagues note that design research progresses probably more by prototypes and frameworks than 'through an academic research agenda with grand theories and questions,' but continue to see this as a deficiency in comparison to more mature disciplines. Although they do not suggest that design will 1 day mature into a science, they leave this door open (Stappers et al. 2014: 170–171). If this is the case, the remaining question is how long would the maturation period be, and how would it happen.

Wisely, they seem to side with the skeptics and give space to a philosophical and sociological ways to understand knowledge. These find supporters in many important design research communities, including empathic and participatory design, and also in the pragmatic tradition that provides the background to Forlizzi's and Zimmerman's work in Pittsburgh. After all, social sciences have a few hundred years of experience, but they have shown little success in becoming anything similar to exact natural sciences like chemistry or physics, philosophical discussions about *Geisteswissenschaften* and hermeneutics cast doubt on whether this is even a valid goal. Skeptical arguments about emulating science have found their way to design literature and they are powerful (Presence Project et al. 2001). Also, designers' self-image as creative professionals leads to skepticism. Professional designers tend to see themselves as creative practitioners rather than problem solvers. If the self-image is one of a creative, then precedents become springboards that have to be redone rather than obeyed, guidelines only guidelines and nothing more, knowledge in fields like ergonomics become checklists and sources of inspiration rather than facts, and so on. If designers see themselves as creatives whose job is to break boundaries, even the laws of nature become malleable constructs. These beliefs are typical in design communities, and given their combined weight, they suggest that underneath the ideal of turning design on a science is like building on sand.

This view poses the question of how knowledge is understood in those fields of design that get close to the social sciences. The most important precedent here comes probably from Scandinavia, where participatory design has been turning to sociology and interpretive philosophy since the eighties, and Helsinki, where user-centered design was built on an empathic basis. Both approaches are interpretive at bottom. Prior to creating concepts, design researchers seek ways to understand how people do things, work, and experience what they are doing. This knowledge is typically generated through fieldwork of some sort, and it is used as a basis for subsequent design activities. The premise is that expert designers have to learn from people, who in turn learn from them. These designers add serious user research to design and see design outcomes as a fusion created in the design process with people they are designing for. Design becomes a social practice of mutual learning. (Ehn 1988, Mattelmäki 2006).

Taken to its extreme, this perspective leads us to see design as a liberal art in the manner of Annie Gentes (2017) who argues that engineering is in fact a branch of the liberal arts. In her view, engineering progresses like the humanities: less by an accumulation of facts and more by a meticulous attention to anomalies as a starting point of research. If she is correct, progress in engineering — not to mention design — comes from making existing things better. Progress is about creating exceptions to things that already exist, and the way to advance knowledge is to take seriously what is already out there to create things that are better suited for the purpose – or at least is different from existing things.

2.6 Sum Up

This chapter has studied knowledge in the field of constructive design research, its relevance to professional practice and the need of research contribution to adopt to standards in order to be exchangeable to other fields of research contributing to and being affected by the practices of design. The chapter builds the description of the designer as from the historical and, in popular media to some extend still appraised, gifted individual artist to the designer as social sense maker, initiator and facilitator of processes of mutual learning in the pursuit of democratic influence on inventions. The argument distills four epistemic traditions of constructive design research where knowledge is generated from: (1) *experiential* insights and objects resulting from design processes, (2) *programs* and theoretical frameworks as foundations of design decisions, (3) *methodically* through scientific measures that help to assess the actual quality and validity of the design intention and finally (4) a *dialectic* constructing of mutual learning between experts and amateurs. These epistemic traditions will be unpacked in the following chapter.

Chapter 3
Drifting in Four Epistemic Traditions

The previous chapter mapped some of the most important characteristics in which constructive design research differs from professional practices and conditions for design. Our focus was on the implications of a shift from design practice to a knowledge-based discipline. We paid attention specifically to how this shift has been interpreted in constructive design research, and how this shift changes the outcomes of design. We saw a wide variation on perspectives from those focusing on how artifacts carry knowledge to those authors who want to turn design into a science. Although constructive design research focuses on artifacts and cannot exist without them, it differs radically from design practice in one respect, which is the context in which claims are justified. In design practice, the context of justification are the design world and the market. In constructive design research, the context of justification is knowledge and design research community.

This chapter build on this analysis but focuses on the lead concept of this book, drifting, and what happens to it when the underlying interpretation of knowledge changes. We focus first on understanding how drifting would relate to various perspectives described on knowledge described in Chap. 2, and then study an underlying variable behind these relationships. This underlying variable comes from our previous work on design accountability (Koskinen and Krogh 2015, and it builds partly on Michael Lynch's (1993) sociological interpretation of science as practical activity. It is worth to note that the epistemic traditions gives various prominence to drifting. For example the *experiential* tradition encourage drifting throughout all phases of constructive design research, while the traditions build in programmatic clarity and scientific compliance are recognizable for drifting primarily in the early phases of research. This is echoed in the balance of using the terms 'drifting' and 'knowledge' when describing each epistemic tradition.

A good starting point of our discussion is again design as a personal construct. In the first sight, drifting is a non-issue in this view. Taken to the extreme, if any decision is up to the designer, he does not need to justify drifting either; he just changes course in his design, redoes the design, and if the outcome is fine, there is no need to justify neither decisions that went into it, nor changes that follow. It is enough for

P. G. Krogh, I. Koskinen, *Drifting by Intention*, Design Research Foundations,
https://doi.org/10.1007/978-3-030-37896-7_3

a designer to say that decisions are his, and there is no need for further analysis. There is no need to be methodic, articulate about how knowledge is embedded in an artifact, nor is there a need for a framework that details the meaning of the artifact. The argument is familiar from the art world, and from popular culture in which it is perfectly fine to tell that creativity is a secret. By implication, there is no need to account for drifting either.

If we look at the implications of the person-centered view to drifting purely in logical terms, the view means that the notion of drifting is trivial. Changing course in the middle of an evolving design has no need to be grounded in anything but the designer's decision. He/she can also just tell about the change retrospectively, locate the drift to something in his/her mind or work process. There is no way to contest this account. The designer can just bluntly say it is his design and nobody else has authority over it.

This is not the only belief about drifting, for sure. For example, design has also be seen as a methodic practice that has its ways of evaluating what is a contribution and what are the proper ways to do things. Designers are accountable to their peers, just like members in any other occupation and profession. There are also organizational accountabilities. If design is done in a company, it must use the methods others in the company use, and not only the outcome but also the process of making needs to make sense to its other members. In this context, designers need to use the tools of the organization, follow its processes, and use patterns of reasoning others can understand. If their work deviates from the ways of the organization, other members of the organization may not understand design decisions, and may sanction the designer.

The hypothesis of this chapter, then, is that drifting has a close relationship to how knowledge is understood. With this hypothesis in mind, we can identify four belief systems about how knowledge drives drifting in design. We call these belief systems 'epistemic traditions' to underline their historical character. The first builds on a long tradition in art and design. It has found a clear protagonist in the *Presence Project* (Gaver 2001), and is shared widely in art and design schools and design consultancies all over the world — we term this the *experiential* epistemic tradition. Another tradition is *methodic* — thus its name. It is familiar in engineering and it has found protagonists from several research communities since the fifties. The third tradition is more recent and builds on the idea that knowledge resides in research discourse and framed by programs, which also drives drifting — we call this *programmatic*. Finally, we describe a fourth tradition in which drifting stems from mutual learning — this we term *dialectic*.

3.1 Experiential Tradition and Drifting

Constructive design research can first be seen in terms of design practice. This epistemic tradition is prominent in the documentation of Presence Project referenced in Chap. 2 and later in Bill Gaver's paper *What should we expect from Research through Design?* In this paper, he tells that 'theory underspecifies design' and argues that design is different from a Popperian definition of science in that design has a '…tendency to make generative statements rather than falsifiable ones.' (Gaver 2012). It is however Kristina Niedderer, who has strongly argued for accepting practical experience as a source of knowledge (see for example Niedderer and Roworth-Stokes 2007).

Looking at the corpus of this book this approach is echoed in among others the work of designers and writers like Jayne Wallace (Wallace 2007) who stay close to traditional design world and drift mainly within the design process. In her PhD work she tries to articulate a way to give more freedom to design decisions that are hard to capture in rational terms. For her, drifting is a matter of a design process and people engaged in it. Her specific case is jewelry designed for individuals. The attempt of the work is to demonstrate that contemporary jewelry design, including the treatment of digital technologies may result in meaningful objects. If the process tells designers to drift, a drift is justified. If we follow her, the value and credibility of the work is not the documentation of design ideas and skills in the process, but the capacity to link information collected in the process, in this case the adoption and redesign of cultural probes she calls 'stimuli' (Gaver 1999). The evaluation of the knowledge created is the capacity of the objects to resonate with its prospective users. It is not in any specific experiential manner determined by the designer, but by the potential of jewelry to find meaningful ways to enter into the lifeworld of the individual people.

The design sensitivity and relevance of the contemporary jewelry designer is further supported by the personal coding (Fig. 3.1) of interviews after the recipients have had a change to live with Wallace's objects (she calls them 'pieces') for about onemonth. She says about the research reasons for giving the objects to the people they were designed for:

> I use the term *hypothesis* metaphorically here; the pieces acted as propositions in the manner of a hypothesis, the existence of which begged testing in interaction with the particular people the pieces were made for. (Wallace 2007: 147)

The argument is that if such designed objects may do this, contemporary jewelry design has valuable competences to offer the design of digital objects.

Another example from our corpus that adopts the argument of an experience being of value because the objects it produces are appreciated comes from a dissertation published by Aarhus School of Architecture in DK, the author, Mo Michelsen Stochholm Krag (2017), who calls his approach 'a counter-practice.' This practice bares the character of interventions in the shape of artistic reshaping of abandoned vernacular buildings in rural village areas to serve the purpose of pointing to an alternative foundation for discussing: decay, preservation, demolition, and transformation beyond measures of economy, identity, and classical guidelines of restoration.

The background for this, in short, is the intensified move of people, that may be witnessed in most countries globally, away from rural village areas and into larger cities, rapidly eroding the notions of liveability that used to define such areas. When

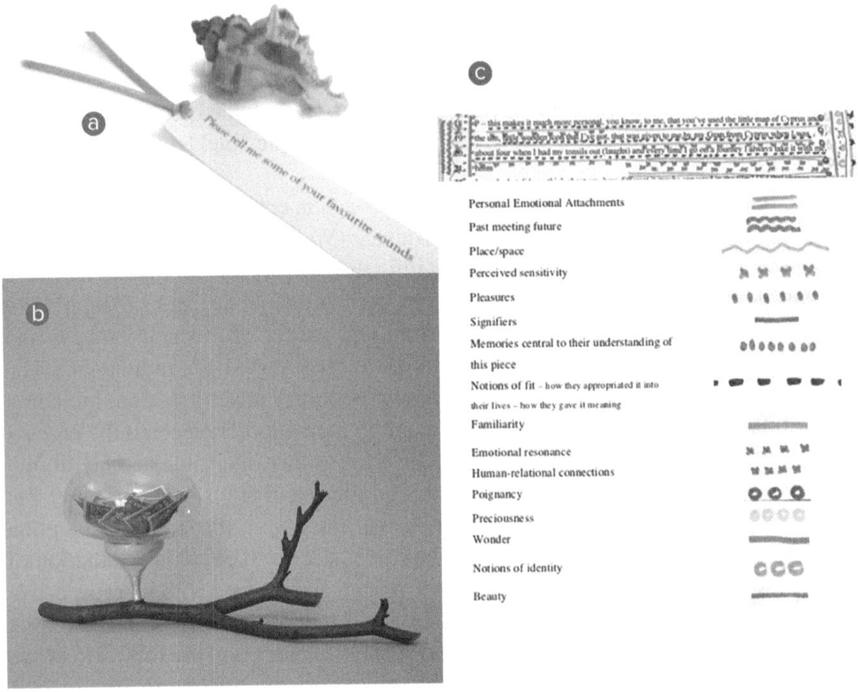

Fig. 3.1 (**a**) Sample of Stimuli, (**b**) The jewelry piece: Blossom, (**c**) Authors coding of interview statements

older buildings are left, rapid decay follows. When institutions (e.g. schools, public authorities, shops) are left, the places are not just empty of function, but also become reminders of the type of life that is no longer supported. This abandonment and decay can be measured in the weakening of local economy as assessed by classical guidelines in terms of the loss of aesthetic quality and cultural disillusionment. Using the same indicators and guidelines the government in Denmark financially supports the demolitions of abandoned buildings, a downward spiral seems inevitable, argues Krag.

A key element on the PhD work of Krag is a practice-informed approach exploring an aesthetically informed strategy to transformation. A significant part of the dissertation is based on cuts made to buildings with power tools by Krag and students of architecture. The work takes place in Thisted, which is one of the most depopulated municipalities in Denmark (Fig. 3.2).

The strategy merges the need to educate students of architecture in the actual technical structure of buildings with creating artistic statements that allows local inhabitants to re-evaluate a building prior to demolition. These interventions are termed as physical dissections. They are explored and slightly modified over the cause of several iterations termed 'generations.' The interventions produce cuts that resemble classical 'intersections' and 'building biopsies' which are exhibited out of

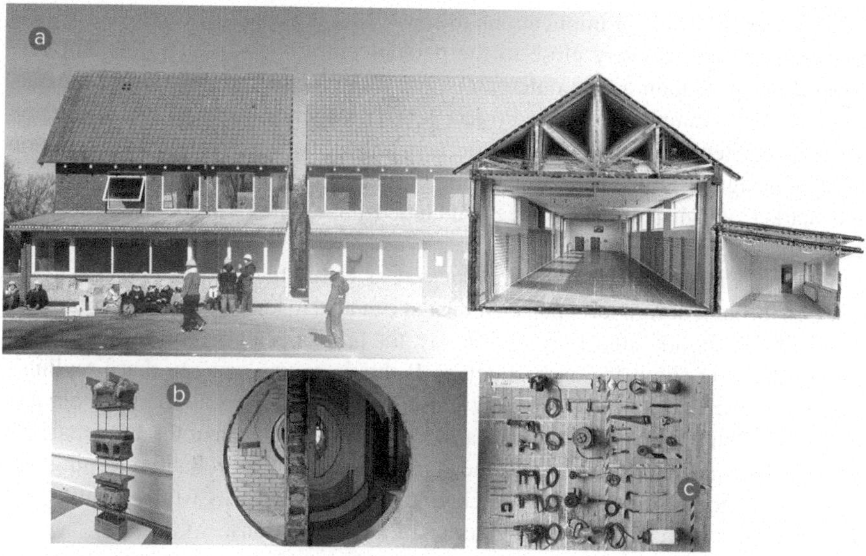

Fig. 3.2 Pictures from Krag's PhD thesis: (a) The physical sectioning of a school, (b) biopsy, (c) powertools

context in a school of architecture. Beyond the purpose of educating students of architecture the interventions also are attempted to articulate intangible cultural values not to decision makers but to the residents of the village. During the performance of the interventions the site and building subject to dissection becomes a meeting place for the villagers to share stories about former inhabitants, the meaning of the place, and the historical ramification of its activities. By reference to the values identified in the UNESCO 'Conventions for safeguarding Intangible Cultural Heritage', the capacity of the interventions to prompt such narratives is used as a support for the claimed quality of the dissection. As may be read from above drifting happens at least three levels: The capacity of the students in dialogue with the physical state of the building conditions the dissections and the actual activities are changed over time framed by Krag as 'generations' of interventions. And finally, the perceptions of the dissections as expressed by local people and stakeholders, both affect the actual cuts and the potential stories to be told.

What connects the two examples are the way that the subject of design directs the course of action and becomes the key motivator for drifting. The produced objects and interventions become a material hypothesis of the claims made, and carried with the ambition of approval and recognition by the people engaging with research. A key element in this is that debate is brought back to the people that are the key recipients of these statements. The correspondence between statements from people experiencing the design production and the character of the hypothesis completes the statement that drifting has occurred. It is the resonance and effect of their artistically inspired artefacts in harmony with the people that becomes the measure of their research. Produced artefacts become the evidence of the knowledge produced.

This view has radical implications to how we understand drifting. As a logical argument, it takes us very close to the person-centric view, which does not care about drifting as long as the outcome is good. In practical terms, this is not true, however. For example, Presence Project (2001), which introduced the distinction between epistemological and aesthetic accountability, was designed to question whether design has to be laid on scientific grounds. Yet, the tenor of the project was anti-market, and its research process included phases that can be seen either as design or as research.

Presence Project is relevant in another way as well. Frayling's (1993) concept of research through design left open how design can lead to knowledge. Because it was done at RCA slightly after Frayling's essay, the project is a good proxy about what he was thinking about the early nineties. If this is so, we can say that Frayling's concept aligned design research very closely with design — as the term already suggests — but it remains ambiguous about whether it is design or research that gets priority. The question about which one is more important is probably meaningless, and best left to be decided by researchers themselves, but it is worth pointing out that Frayling's term is ambiguous in terms of its implications.

3.2 Methodic Tradition and Drifting

A polar opposite to the art and craft view has been expressed in design engineering literature. In terms of its implications to drifting, the methodic view of Blessing and Chakrabarti (DRM 2009) — and their predecessors in the Ulm School of design, in design methods movement of the sixties, and successors most recently under the Delft Method — could hardly be more different from the traditional, person-centered notion that gives a designer authority over drifting.

The *methodic* epistemic tradition, as we term it, in fact goes to the other end by locating authority over design to knowledge produced by those methodic practices that define the shape of the design process. In doing so, it paints drifting either as an illegitimate practice, black art that needs to be rooted out of a rational design process, or sees it as method variance that has to be controlled. It is a threat to rationality and the rational worldview in the behind this vision: any act of drifting needs to be grounded in methodic knowledge, and any act of drifting has to be justified by this knowledge. Taken to its logical conclusion, design would become like any field of engineering: not quite science, but almost. When its methods are perfected, they become vehicles for creating knowledge and then eliminates hunches and personal whims from design.

This view, of course, neither rules out novelty nor turns designers into pawns of method. In this view, drifting can happen in many ways. It can be built into the method. The key word is control rather than banning drifting. Drifting can happen, but it has to be controlled. Most typically, this takes place in terms of decision points in the method, or in industrial language, in gates in which designers make decisions about whether or not to take the project further, and in which direction to go. These decisions are primed by the previous phases of the process, and presented as decision alternatives. The process is rational in that any instance of drifting must be subjected to explicit evidence. The process is designed to rule out tacit knowledge and person-dependent hunches of the sort Wallace and Krag propounded.

Drifting can also happen in thinking and the early research phase. Most descriptions of design in engineering depict it as a process in which there is a lot of freedom in the early stage of the process, but very little in the later stages. This reasoning behind this shape is usually grounded in economics.

When looking to our corpus, the methodic epistemic tradition is particularly evident in the Dutch dissertations from the technical universities in Delft and Eindhoven. The production is rich and the dissertation often true to the tradition. There are exceptions though, like Ambra Trotto who in 2011, handed in the thesis *Rights through Making*, which was exploring how processes of design may emphasize the experience and promotion of human rights. In line with the tradition, *Designing for Rich Interaction* was the title of the dissertation of Joep Frens in 2006. Building on theories on affordances, Frens (2006) believes that 'design[ing] for interaction is to design the form-properties of a product that comprise the interface' with particular emphasis on cognitive, perceptual-motor, and emotional skills…' Based on the research interest and potential contribution — the rich interaction framework — a tangible digital camera became the pivoting point of his dissertation. It was tested in the dissertation to learn about the rich interaction framework. The selection of design alternatives and choices are rationally argued, which turns his cameras into hypotheses that could be tested. The three major research activities of the work are performed as scientific experiments:

1. A comparative study of lab based usability experiment including the rich interaction camera and a classical contemporary digital camera design.
2. To remedy the shortcomings of the first experiment, a working prototype of the rich interaction camera including four interface variations was build.
3. Finally, the rich interaction camera was subject to real-world testing by a total of 24 participants in a photo studio where a protocol of Marc Hassenzahl's product quality assessment scales (2004) was deployed.

The test results were coded according to the camera variations tested, protocol questionnaires used and grouped according to which research interest is enlightened by the data collected (Fig. 3.3).

As can be viewed from the above, this is an attempt to rationalize design decision that directed the building of the camera. Drifting is controlled, but still happens. Despite its attempts to be scientific the thesis, however, also builds on creative practices and choices typical to any design process.

One of the virtues of this vision is philosophical in nature. This view would turn even invention in the very beginning of design into a rational exercise. This is probably too ambitious, looking at what philosophers of science regard as mainstream. After Karl Popper's falsificationist philosophy of science, some philosophers have for decades made a distinction between the logic of invention and the context of justification. In the sciences, the latter should be as unambiguous as possible, while the early phase can be inventive. Invention almost by definition has no logic; it becomes the place for creativity. The view is not practical either. Indeed, Overbeeke

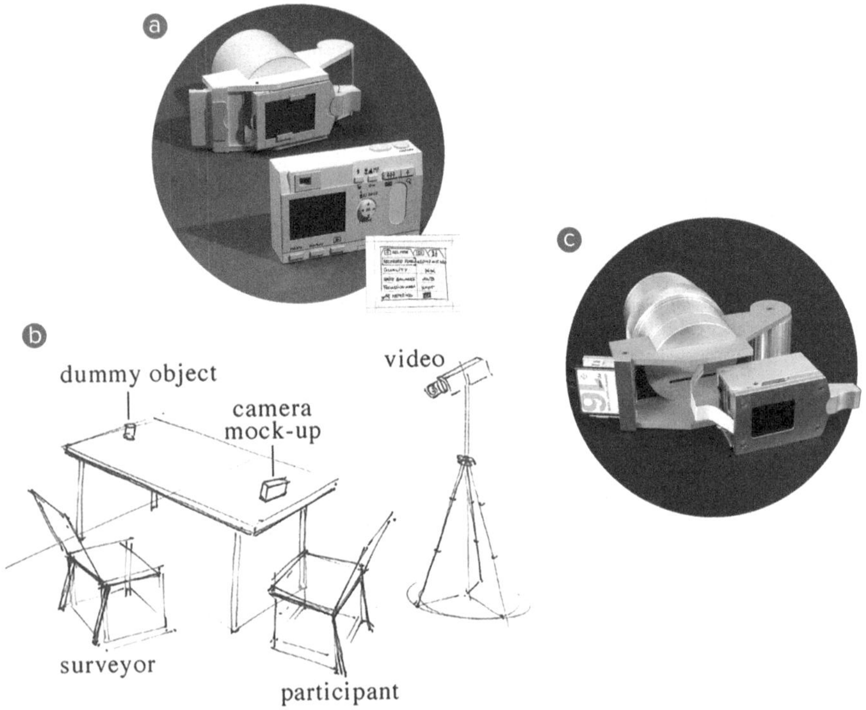

Fig. 3.3 (**a**) Cardboard cameras for experiment 1, (**b**) Lab test set up for experiment 1, (**c**) Final prototype

did not go this far in his own research work, which departed from his vision significantly.

Although this view is probably wrong if taken too far, the worldview it articulates is attractive to those who have grown up in the mathematical worldview of the sciences. This is how the formal sciences work, after all. Inspiration can come from anywhere, but when proving a theorem, mathematicians must follow the rules of their branch of mathematics. The main reason to be sceptical about this model is that as soon as we leave the formal sciences, we tend to see lots of methodic inventions and tricks of the trade that make research work. The belief that method variance, as it is known in statistics, could be eliminated completely, is probably overly ambitious and so unpractical as an ideal that it is better to leave it to elementary textbooks.

3.3 Programmatic Tradition and Drifting

When we turn to the idea that knowledge advances through frameworks, drifting happens in frameworks and research programs — thus the name *programmatic epistemic tradition*. One of the more interesting findings of Chap. 2 was that we saw how the debate about research through design has broken into several lines. While some researchers have followed the message of Frayling and turned their attention to design artefacts, others have turned their attention to the conceptual and theoretical scaffoldings of these artefacts. This has been the line best known from the work of two researchers based in Pittsburgh, Jodi Forlizzi, John Zimmerman, and their colleagues Shelley Evenson and Erik Stolterman, the latter working in Bloomington, Indiana and Umeå, Sweden (Zimmerman et al. 2007, 2010).

This view sees frameworks that illustrate reasoning in design and generalize from it as crucial sites for producing knowledge. In this view, drifting occurs to some extent in design work, but more importantly, it occurs in conceptual work in which researchers create frameworks and theories, and debate their pros and cons. If we think about drifting in this view, it may happen in several ways. It may happen in design projects, but also their conceptual environment. Researchers build on their research predecessors rather than design exemplars, as practitioners would. There can be a lot of ambiguity in design objects, and it is the job of the debate about frameworks to clarify this ambiguity.

The programmatic epistemic tradition is fairly pervasive in the Nordic countries. In the following we exemplify this through an early example Martin Ludvigsen's *Designing for Social Interaction* in 2006 and recently Yiying Wu's *Bicycles and Plants* in 2017. Through a set of design cases most prominently analysed in the perspective of Erving Goffman's (1978) model of social encounters Ludvigsen identifies a framework for Social Interaction and a model for Situational Interaction Mobility. The latter is presented in order to '…describe[s] the change in level of social interaction in the framework.' (Figs. 3.4 and 3.5).

The category of 'the collective user' is his contribution to the framework which depicts how technologies foster particular social relations. A banal example of the below would be: a table is located in a room, everybody knows it is there. However, the table is largely unnoticed (distributed focus). Someone come in and brings to the

Fig. 3.4 Ludvigsen: (**a**) iFloor, (**b**) Media surfaces, (**c**) Star catcher

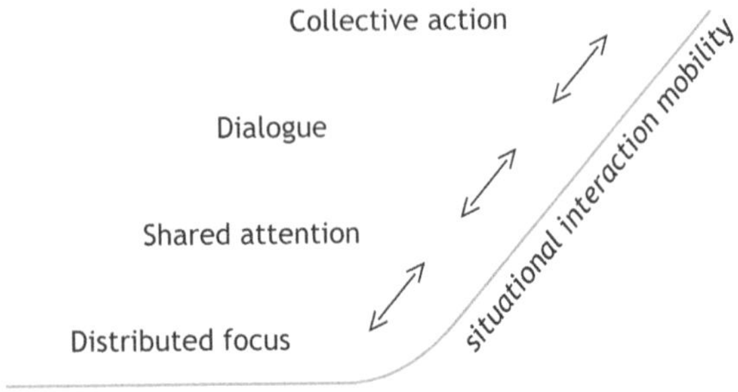

Fig. 3.5 Martin Ludvigsen: social interaction in the framework

Fig. 3.6 Plant hotels in different social contexts: community, gallery and university

attention of all present that the table needs to be moved to the room next door (shared attention). Discussion emerges about what and how to do next (dialogue). Due to the size and shape of the table at least two persons are needed to carry and coordinate actions in order to get the table out the door (collective action). At the time of the research work, little attention had been brought to the qualities of the final category. *iFloor Project* was maybe the first interactive prototype to point to this, and facilitated to infer yet overlooked design opportunities of digital interactive objects. Ludvigsen's participation in a larger research center for Interactive Spaces (2002–2009) in Aarhus, Denmark, allowed him to drift between projects in search of design frames with the best opportunity to inform his research interest on 'How to design for Social Interaction?'

Another, drifting, framework building design researcher is Yiying Wu (2017). Inspired by the framework for Social Innovation (Ezio Manzini 2015) Wu pays particular interest in designing with 'creative communities.' In this perspective, she follows and analyses six international and Finnish cases that serve as inspiration for her final construction of a 'plant hotel' (Fig. 3.6).

In this process Wu drifts in at least two significant ways. First, she changed perception of the designer role — anticipated to be the skilled helper based on recordings of context, but rather than being the expert she became the learner through adoption:

> Hence, instead of using this piece of ethnography to inform design, I positioned it as a mirror to re-examine normally unexamined design… My designerly response was to open a

> workshop of a similar kind by myself… It should be a service setting where people come to work for others. 'Work' refers to the notion of a capable being with a sense of responsibility and a high level of participation. 'For others' refers to intensive social relations and interactions among participants. (Wu 2017: 43)

Second, originally the Plant Hotel was envisioned to be one of three different service design interventions. But the capacity of the concept to probe social relations and amplify a sense of giving and fostering care rendered it to be a strong tool in exploring service design as a social practice instead of its traditional field of study: management and engineering. And her deployment of the concept in five different versions and contexts helped her build a program that reveals how design can be used as a means for anthropological studies (Fig. 3.7). This work shares several overlaps with Joachim Halse (2008) whose thesis *Design Anthropology: Borderland Experiments with Participation, Performance and Situated Intervention* explored some of the same phenomena, but from the perspective of anthropology.

As can be witnessed in the work of both Wu and Ludvigsen contributing to existing frameworks point to these becoming programs for constructive design research. Ultimately, these programs change expectations, perceived standards of design, as well as perceived standards of proof.

At first sight, this sounds antithetical to the credo of Presence Project. What could be more different from claiming that drifting is driven by the myriad of decisions designers make in their studio, and that drifting happens in research world debates? When phrased this way, the views are different indeed. Yet, it may also be argued that these views are in fact quite a bit closer than they seem in the first instance. We see this argument as only one way of capturing drifting, and in fact, this view complements the view of Presence Project. We do not see any significant

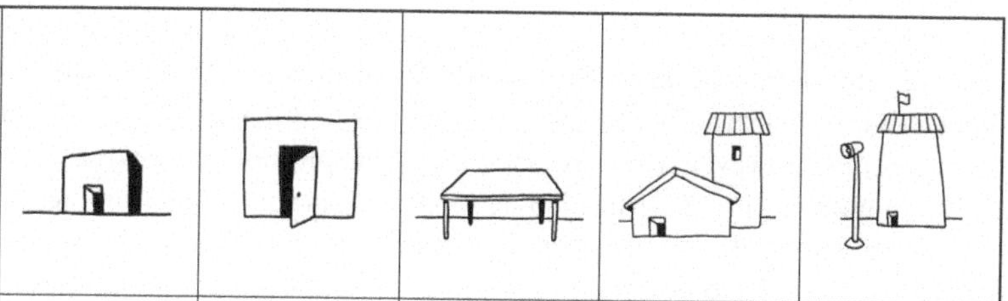

PLANT HOTEL					
Social setting	In a neighbourhood gallery, Helsinki	In front of the professors' office, Helsinki	At an academic conference, Stockholm	In two elderly service centres, Helsinki	At the border of North & South Korea (speculative)
Social relations	How shall we treat our neighbours?	How shall we treat our professors or students?	How shall we treat our guests?	How shall we treat the elderly?	How shall North & South Koreans treat each other?

Fig. 3.7 Framework of how design implementation of a 'plant hotel' probe social relations

conflicts with the idea that drifting is driven by user studies either. It can be a function of user studies, but it can also take place later in a design process.

Looking at the notion of drifting, the idea that objects may remain unchanged while discourse changes sounds illogical. However, it is not necessarily so. In any field of research, there is theoretical debate that tries to find explanations for the field's facts and observations. Observations remain the same, while theory drifts until it settles. Design is no different in this regard, we believe. Explanations given in the first hand by researchers who have done research have a certain degree of authority, but there is usually no need to think their accounts are authoritative. Another reason for thinking that this is in fact normal is that there is no reason to think that artefacts in the center of constructive design research can be emptied in any one account. Many artefacts produced by researchers can be seen in many ways, and if frameworks are how researches make sense of these interpretive possibilities, so be it. In this tradition, discourse drives any field of research probably more than artefacts or observations.

It is also worth pointing out that this view encourages drifting very much in line with of Christopher Frayling. The main difference is that it encourages researcher to give conceptual handles to understand their design — it establishes a program. These handles help other researchers to make sense of design work: where it came from, what was the problem it dealt with, how the solution evolved, and what can be learned from it. Drifting happens in research projects, but also in discourse around them. Frameworks facilitate drifting and are a crucial part of it.

3.4 Dialectic Tradition and Drifting

In line with the interest into the social life world exhibited by Ludvigsen and Wu we now turn the notion of *dialectics* as a driver of drifting. Many of the dissertations included in our corpus involve users in processes of design. Users play many roles in the corpus, however: they are a source of inspiration, co-designers, evaluators, and even guinea-pigs. This has led us to shed a particular light on the dissertations that are driven by the involvement and change impact that the prospective user of design artefacts. Here designers work in a community, encourage these communities to voice their views, and in a dialectical manner help the community to surpass its initial starting points. The process runs from theses through antitheses to syntheses, but in design almost always without a philosophy of history. The main exception may be activity theory, which builds on Vygotskian Russian psychology in debt to Marx (see Kuutti 1996).

Previously, we have briefly talked about empathic and participatory design. We argued that they each introduce a different kind of idea of how knowledge is construed and how knowledge conditions design. The heart of the issue is that when including and studying people in research, knowledge will almost never be precise. The main exceptions are areas such as demographics and movement between cities. The reasons for imprecision are familiar from the social sciences: correlations are weak; there are so many intercorrelations that any explanation can always be contested because it always neglects some background variables; people react to their circumstances and can choose their line of action; and ethical reasons make controlled experiments in the scientific style impossible.[1]

Knowing this, many designers have drawn the conclusions. They are happy with creating knowledge they need in participating with people, and they do not claim that their knowledge is authoritative. As indicated above, there are significant differences between empathic and participatory design. To unpack them, we study two more Scandinavian constructive design dissertations to illustrate how drifting differs in them.

The Helsinki-based designer and professor Tuuli Mattelmäki (2006) sees her craft in dialogical terms. For her, design ideas are meant for participation, where they are supposed to live for the duration of her research projects. The knowledge she creates is for her not an unshakeable foundation built on knowledge, but a temporary base that helps design forward. She also rejects the idea that designers should be flies on the wall, observing people: design is about change, and designers are change agents, and she does not want to betray her tribe by telling their traditional ways are somehow flawed only because she sees herself as a researcher. Her research methods are meant to serve as conversational props that prime people to think about issues she finds relevant to her design work.

These convictions are evident in her language. She described her approach as empathic, and sees it as a form of co-design, and she sees herself as a facilitator rather than an expert, who knows what people need better than the people themselves (Fig. 3.8).

Christian Dindler (2010) delivers insights on how institutions such as museum and archives may move their conception of exhibitions concepts from the perspective of visitors as passive recipients of experiences to active participants who co-construct engaging experiences. He does so by narrating what he terms 'Fictional spaces', in which participants may co-imagine acts and connections of the exhibited artefacts. The objective of the design activity is thus not an artefact, but the capacity to involve participants and change their perception of what an engaging museum may entail.

In the early days of Participatory Design in Scandinavia — then called Cooperative Design — the aim of such processes was to develop technology and

---

[1]This has been said particularly colorfully by Bruno Latour and Steve Woolgar in their book Laboratory Life from 1979: 'Occasionally, when members of the laboratory derided the relative weakness and fragility of the observer's data, the observer pointed out the extent of the imbalance between the resources which the two parties enjoyed. 'In order to redress this imbalance, we would require about a hundred observers of this one setting, each with the same power over their subjects as you have over your animals. In other words, we should have TV monitoring in each office; we should be able to bug the phones and the desks; we should have complete freedom to take EEGs; and we would reserve the right to chop off participants' heads when internal examination was necessary. With this kind of freedom, we could produce hard data.' Inevitably, these kinds of remarks sent participants scurrying off to their assay rooms, muttering darkly about the 'Big Brother' in their midst.' (Latour and Woolgar 1986: 256–257)

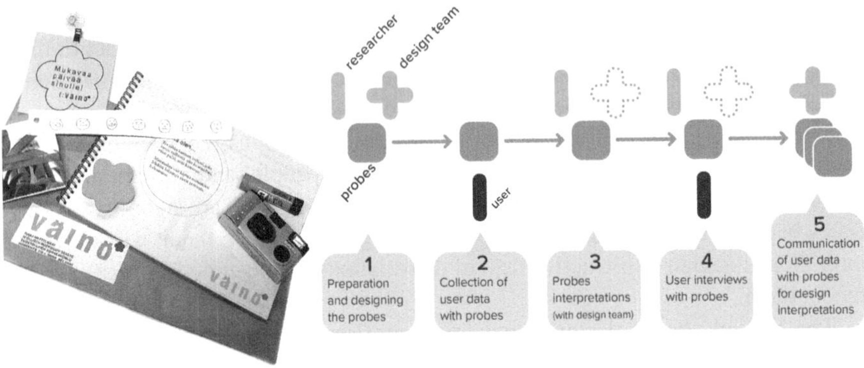

Fig. 3.8 Collection of probes and model of analysis as applied in Mattelmäkis dissertation

systems design that would enable *quality* of the technologies, *emancipate* the workers and *democratize* the organisation of work (Ehn 1988). The focus of design quickly moved from software and systems development to educating workers on the properties and consequences of digital technologies. While this academic heritage runs as an undercurrent in Dindler's dissertation, it became evident in a later project (Dindler et al. 2011). A regional museum of archaeology in Denmark approached him with two concerns: the number of visitors to the museum were decreasing; and due to Danish treasure regulations their most valuable artefact — a mummified bronze-age woman — had been transferred to the National Museum. To remedy this, they had applied for and been granted funds to install a digital (VR) version of the woman.

Rather than immediately jumping on to the bandwagon to produce the requested installation, Dindler created a counter-brief and opted for a *dialectic* participatory design process. Already here the project drifted: this move brought tensions among personnel to the surface. These tensions were about who got the idea, who applied for the funding, who wanted something else, and so on. When settling such disagreements through co-design workshops, a further drift in focus enabled Dindler to see that what might unite all museum personnel was the aspiration to become a regional center for archaeology rather than a collection of artefacts. The installation designed in the project became the temporary materialization of this potential future. (Fig. 3.9).

This process shows how the value for the personnel was not the installation, but the opportunity to re-orient the organization from a keeper of culture to a research hub. Studies like (Bossen et al. 2010) similarly argue that the real value of participatory design is not always the influence on the actual design, but equally often the human and professional competencies build during such processes — the *dialectics*.

Mattelmäki builds her approach on a long history of user-centeredcentered design, and she has occasionally also been publishing her work in participatory design conferences. She has also been working with several Milanese designers, whose approach has shifted from scientific and systemic approaches to a human and

Fig. 3.9 An interactive burial mound, where digital images of treasures are projected on to the dirt as by tracking technologies are they revealed through removal of dirt

social direction. Their word of choice has, like Mattelmäki's, been dialogue (Manzini 2015 on cores theme). Just like the early participatory designers were concerned with fostering change and allows (all) participants to knowingly transcend current social practices, this has been true of Mattelmäki's work as well.

All this is well in line with interpretive social research, which tends to see its outcomes not as authoritative statements of knowledge. For example, the sociology of professions has since the sixties been fascinated about well-informed middle-class patients who come to the doctor's office ready to tell what is wrong with them and have difficulties in accepting the diagnosis. In science studies, the fall of authoritative knowledge has been well captured by the sociologist Helga Nowotny, who distinguished Mode 1 and Mode 2 knowledge. Mode 1 knowledge refers to the era of big science fueled by cold war, where science gained a largely autonomous role as long as it could deliver the weapons. Mode 2 in contrast describes an era in which knowledge becomes negotiable, and has to be robust rather than valid to survive in public discourse.

In the dialectic epistemic tradition, beliefs about knowledge get grounded in the community in which knowledge is created. Knowledge has to be robust to survive in conversation, but the aim is not to create knowledge claims that would survive years or even centuries. These beliefs have profound consequences to how drifting is supposed to happen. Drifting happens if people in the community say so and if a designer like Mattelmäki finds this fruitful for her project. Design ideas have to survive the dialogue, and designers need to get prepared to respond to surprising, ill-informed, combative, and even mean arguments targeted at their design. They also must safeguard against giving people a right to veto their designs.

All this detaches design from strong knowledge claims, and pushes the source of change to designer-people dialogues. Unlike in earlier versions of user-centered design, designers are not required to conduct careful studies that lead to valid knowledge that then informs design. Rather, they need to study people, work with them in a co-design fashion, listen to them carefully, and react to what they are being told. Knowledge is a secondary concern to the human connection — the dialectic character of the tradition.

While having listed and argued for the traditions above there are hybrids, like Otto von Busch 2008 dissertation which are impossible to describe unambiguously according to either of the four epistemic traditions. Although most of the dissertations does include elements from more than one tradition, they often lean towards one of them.

3.5 Drifting, Accountability, and Context: The Art of Being Robust

This, and prior, chapters has made several references to the 'context' of research. We have used the term 'context' on purpose because we believe that understanding any practice (and in our case, design practices in line with epistemic traditions) requires paying attention to where it comes from and where it is intended to contribute to. This is a lesson we have learned from sociology of science and technology, which has repeatedly shown that scientific statements — and just as any other statements — get their meaning in context. For example, the concept of 'mass' has a definition in physics, another definition in cancer research, and a third in communication studies. It would be worthless to argue which is the best; all have value in their own right. This argument has been expressed in design literature by Jung-Joo Lee (2012), who claims that the success of cross-cultural design methods depend on the practical base of these methods. She argued against the idea that research methods are universal and portable from one culture to another.

Although these are philosophical and sociological discussions meant to answer questions in philosophy and sociology, they have informed our thinking. For us, they tell that to properly understand constructive design research, we need to study it in those contexts in which it is being made. What is knowledge in one context may not be it in another. In turn, this means that we can understand better how research works. Some of the apparent incompatibilities we see in design research disappear when we put research in context; we may also see how some apparent agreements turn into incompatibilities. For us, this context is partly practical, consisting of things like instruments, measurement conventions, and notions of proof. It is equally much intellectual, consisting of those discourses in which findings are framed and debated, contested and accepted.

Our way of probing the contexts builds on our work on *design accountability* (Koskinen and Krogh 2015). We have argued elsewhere that design research is accountable to design practice and design research alike, and that solid research either acknowledges the boundaries set by these environments, or risks being labelled as irrelevant. This will be a recurring theme in the following three chapters and in particular Chap. 4. If we are correct, design researchers talk to multiple audiences simultaneously and have to find a way to select who they want to be accountable to. Even though we do believe in the importance of design accountability, we also know it has limits in research. Taking it too far would push design research back to design, which would not be fruitful in a discipline that tries to build a research practice and a discourse it may share with other disciplines.

If this view is correct, it tells about where to find some of the limits of drifting, and how far it can go. The room for drifting can be limited if researchers work in an engineering-oriented environment that tend to stress methodical rigor in ways that would not be appreciated in design based cultures that put premium on contribution and relevance, or art-oriented environments that expect surprises and controversy spurring imagination.

How far can drifting go amidst these surrounding beliefs? To full anarchism in which everything is open all the time because researchers can freely choose their context? The answer to this question is a clear no. Even when researchers like Wensveen start their studies from cultural probing and finally end up testing their theories with statistics, they respect the rules of each part of their study. What is regarded as knowledge — or acceptable and interesting — varies, but if research wants to be good, it has to be on par with the best practices in each of its subsections. Importantly, accountability sets limits to drifting: if design research is aesthetically accountable and drifts freely, it has to be able to face criticism from practicing designers, who have their own terminology and way to define their work (as Anna Valtonen 2007 has suggested). If it is epistemologically accountable, it has to survive the criticism of fellow researchers.

It is good to keep in mind that although we believe that to understand constructive design research, we need to look at it in its context of production, we do not believe that this context in any way explains what researchers do. This would be a simplistic, causal argument about the relationship between action and context. Rather, we think that beliefs about the context and the role of knowledge in it are built into research. After researchers have defined a relevant context, they also construct beliefs about who they are they accountable to. The context is *made* relevant and consequential in debate; constructive design researchers choose their paths and act on values dear to them. The values come from four main traditions we have described: the artistic quality and spurring imagination – *experiential*, the methodological strictness and enabling predictability – *methodic*, the theories, frameworks and advancing programs – *programmatic*, and the activist result changing participants perception and enactment of everyday life – *dialectic*. The three following chapters dive into more detailed activities for creating knowledge claims and analyzes drifting in them. They unpack how hypothesizing (Chap. 4), experimenting (Chap. 5) and evaluation (Chap. 6) happens in constructive design research, and how the interest of both serving knowledge production and relevance causes drift. The pivoting point for the following Chap. 4 is the presentation and detailed description of the Knowledge-Relevance model.

Chapter 4
Design Hypothesis: Knowledge-Relevance Model

This chapter asks how constructive design researchers construe hypotheses. What guides the construal process, is it theory, research program, or design reasoning? We will argue that this question involves two sub-questions: one concerns the hypothesis that guides research, another the hypothesis that guides design. We argue that constructive design research has to balance both of these to be effective, and we also point out that many of the controversies we have discussed in the previous chapters are in fact efforts to grapple with the Janus-faced character of the discipline. It is this character that has inspired us to suggest the Knowledge-Relevance model that help the constructive design researcher to balance the core research activities when the process is driven by design.

The word hypothesis has made occasional entries into design research, mostly in the writings of Caroline Hummels (2000), Overbeeke et al. (2006) and Stappers (2007). We define a hypothesis as usual as an explanation based on theory. In design, obviously, theory is a loose concept and few theories are as precise as in the exact natural sciences. Theories in design function more like in the social sciences, where they are understood as a guiding light — or a sensitizing device — rather than provable or refutable facts. In constructive design research hypotheses take several forms ranging from graphic depictions, products/ spaces, services and systems (Buchanan 2001), and even social forms like communities (Anna Meroni 2017). In practical terms, when design researchers build hypotheses, they function the same way as any researcher would. They start with whatever hunch they have, do research in the libraries (for theory and literature) and in the market (for products), and start to rule out some alternatives, building a product that exemplifies their initial idea. This process leads to the formulation of more precise research questions that keep only some of the initial ideas alive.

At some point, research gets precise enough for constructing a design artefact that is then subjected to scrutinizing. The important point to note is that at this stage, there is a world of difference to the initial hunch: the process has created a theoretical world that gives meaning to the artefact. It is not the world of the layman anymore; it is a well-informed and precise world that builds on two several classes of

P. G. Krogh, I. Koskinen, *Drifting by Intention*, Design Research Foundations,
https://doi.org/10.1007/978-3-030-37896-7_4

sources, including theory and literature, existing products, and the acts and practices of designing.

As this formulation suggests, just like design, design research is deeply embedded in what already exists (see also Mitcham 1986). Whatever they do, they have to confront the history of design and design research – not only in the beginning of their work, but all through it. Design, as Armand Hatchuel's C/K theory (conception-knowledge) argues, always proceed as an interplay between novel concepts and existing knowledge (Hatchuel and Weil 2008, 2011), and this is true of research as well, we believe. This embeddedness is something many of the accounts of design are missing: they overstate the creative qualities of design, and forget the fact that creativity almost by definition builds on what already exists. As the research process goes on, it keeps making new things relevant, and incorporates some of these into the evolving composition. As our knowledge-relevance model suggests, many of the excursions out of knowledge bring researchers in contact with cultural, social and technological stakeholders.

4.1 Mapping Design Practice in Relation to Knowledge

There is a rich tradition in engineering and constructive design research to propose development models in the attempt of streamlining the design processes. The models developed for this book complement to existing models, and the Knowledge-Relevance model will in particular provide means to map out relations between practical design activities and existing theoretical knowledge. The reasons for models in design practice and design research are numerous: facilitating collaboration with others, ensuring soundness in decision making, ability to deliver on time etc. Figure 4.1 depicts four samples of characteristic development models in design.

Both the waterfall model and the iterative model are articulated in software engineering and development. The fuzzy front-end is well-known in product design pointing to the (often) need of challenging the brief and the question at hand in greater depth. The model is the authors illustration of the way of working promoted

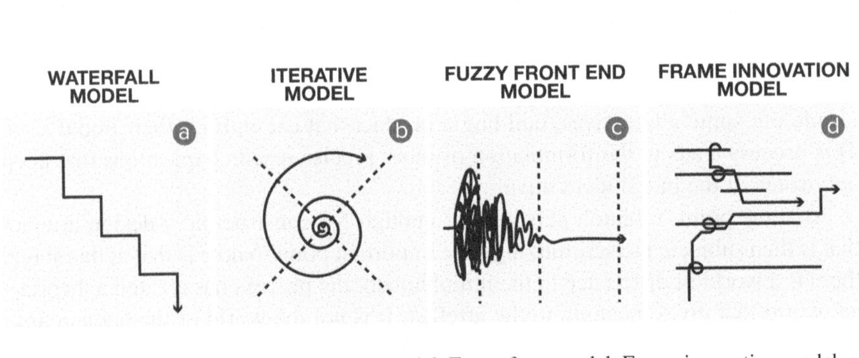

Fig. 4.1 The waterfall model, the iterative model, Fuzzy front model, Frame innovation model

by Dorst in his book *Frame Innovation* (Dorst 2015) where several ideas are kept on the table and all evaluated in relation to a variety of frames rather than assessed according to a particular design challenge. This is complementary to the underlying assumption of earlier models. In them, research processes are guided by a desire to converge to produce a piece of design.

Figure 4.2 attempts to depict the cognitive models of the practicing designer and shows how knowledge is related to the process. Nigel Cross (2018) is interested in illustrating the loop of analysis of synthesizing processes in (product) design. Stappers (2007) sees a designer as someone who scrambles together a set of ideas from a variety of places, and puts them together into a spiral-like form that ultimately leads to an outcome, which typically is a product. In his model, the product becomes a physical hypothesis. The red arrows describe knowledge entering and being disseminated in the process.

Other design models either adopt a language key to design but mostly unfamiliar to disciplines outside design (Binder and Redström 2006), or see design as a hybrid discipline build on theories, methods and evaluation schemes adopted from other disciplines (Zimmerman et al. 2007).

By reading through the corpus, we saw that constructive design research today has increasingly clear contours and contents outlining it to be a research discipline on its own right, however. As noted in Koskinen and Krogh (2015) constructive design researchers hold themselves accountable to design practice before any field of research outside design. Their aim is to be taken seriously by practitioners, but also ground their work on theory, whether ecological psychology (for example, Wensveen 2005), phenomenology (Deckers et al. 2013), symbolic interactionism (Katja Battarbee 2004; Ludvigsen 2006), or Situationism (Presence Project et al. 2001). They aim at contributing to theory and practice alike.

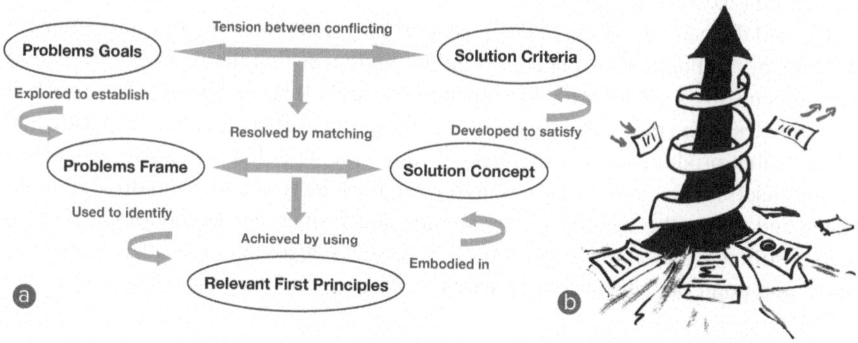

Fig. 4.2 Cross, model of designer working, Stappers 'Arrow' model

4.2 The Knowledge-Relevance (K-R) Model

As we have already argued, constructive design research is not a theoretical discipline that aims at describing and explaining the world, but also has a practical, constructive component. In the following we present the Knowledge-Relevance model in which the act of designing — the *design experiment* — is at the heart of the model. Along with *hypothesizing,* and *evaluating, design experimentation* is the core activity of what Donald Schön (1983) once described as an evolving 'conversation with the material'. The prominence we give to *Hypothesizing*, *Design Experimentation* and *Evaluation* is reflected in that these are key concepts of Chaps. 4, 5, and 6 of this book. In our view these fundamental concepts of constructive design research are continuously brought into play in a dialectic process where the designer externalizes and manifests ideas, hunches, and concerns that 'talk back' to both herself and potential collaborators and thus helps constructing increasingly refined and thoughtful products. What is further significant to constructive design research is that this dialectics serves both *knowledge* production and *relevance*. The measure of *knowledge* creation is contribution to *theory* and the measure of *relevance* is *impact* i.e. the fitness and ramifications of the design experiment — and along the lines presented in Chap. 3, to whom and what constructive design research holds its self accountable to. The graphics of *knowledge* and *relevance* in Fig 4.3 indicate that a constructive design research project produce particular and limited contributions to basic underlying *theory* and *impact*. The proposed model may help both to frame processes of sketching, schemes of understanding full-fledged design proposals, and at scale of research project. The purpose of the Knowledge-Relevance model is to enable the constructive design researcher to keep track, model and facilitate discussion if the balance of research activities and concerns is prober and in accordance with the research pursuit. The model presented in this book builds on a prior version[1] informed and complemented by the work of Koskinen and Krogh (2015) (Fig. 4.3).

We will deepen the notion of design experiment and methods of experimentation in Chap. 5. At this point we need to remind the reader that the word *experiment* in our language does not refer to experimental research. Instead, we want to stress that although designers produce design objects, these can also be treated as experimental artefacts that produce knowledge. Seen through the model, a *hypothesis* articulates the premise(s) under which any research work must be read and understood. It articulates and limits the validity of the studies and frames the methodological landscape. Following this, the *evaluation* is a detailed accounts of what is subject to study, and point out appropriate research techniques and possible outcomes.

[1]The model is a reworking of a model originally developed in cooperation with T Markussen, M Ludvigsen and A.L. Bang (Bang et al. 2012 as a part of the documentation of three doctoral courses hosted in Denmark between 2012 and 2014 also reported in (Krogh et al. 2015, Markussen et al. 2015).

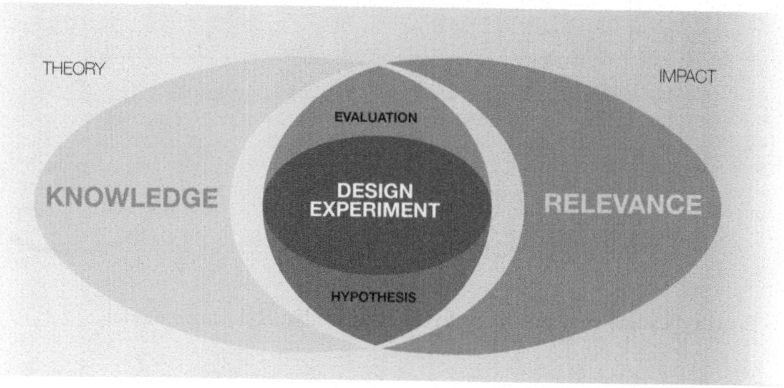

Fig. 4.3 The knowledge-relevance model

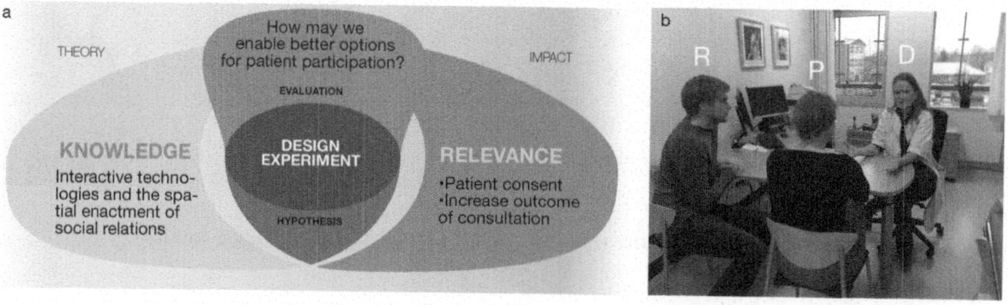

Fig. 4.4 Contextual interview and how it maps to the Knowledge-Relevance model

Against the backdrop of *theory* and *impact* at either side of its vertical axis the model points to the double concerns and contributions of constructive design research: *relevance* and *knowledge*. These are driven by different measures of criteria and incentives, and produce results of different value associated with either *theory* or *impact*. It is this duality that is key to constructive design research. In accordance with the activities of the central vertical axis in the model constructive design researchers oscillates between modes of analysis and synthesize. When doing so the focus of relevance and building of knowledge often inclines drift in objectives, means and measures of evaluation against the backdrop of *theory* and *impact*. In the context of literature we have dealt with earlier, the controversy about constructive design research often stems from such diversity. While some authors primarily bind design experiments to the artefact, others bind them to theory, research programs, and debate.

Vignette: an illustration of the model

We can illustrate the Knowledge-Relevance model as mapping tool by the work of Thomsen and Schnedler (2017 (master thesis); Thomsen et al. 2018). In Spring 2017, Thomsen and Schnedler found interest in the work of (Krogh et al. 2017) on how interactive technology conditions our enactment of social relations (Figs. 4.4, 4.5 and 4.6).

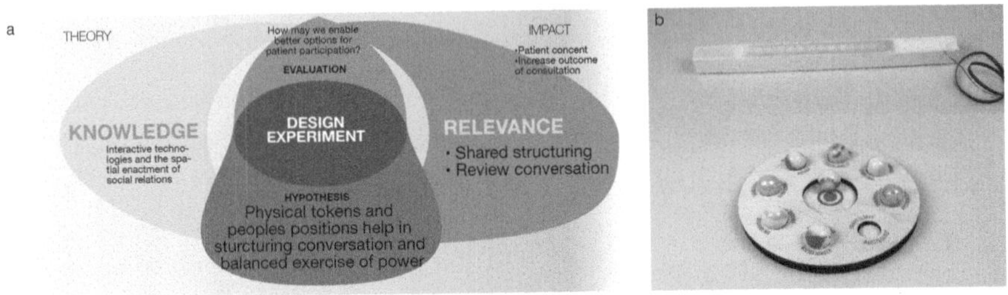

Fig. 4.5 Provotypes and how this maps to the Knowledge-Relevance model

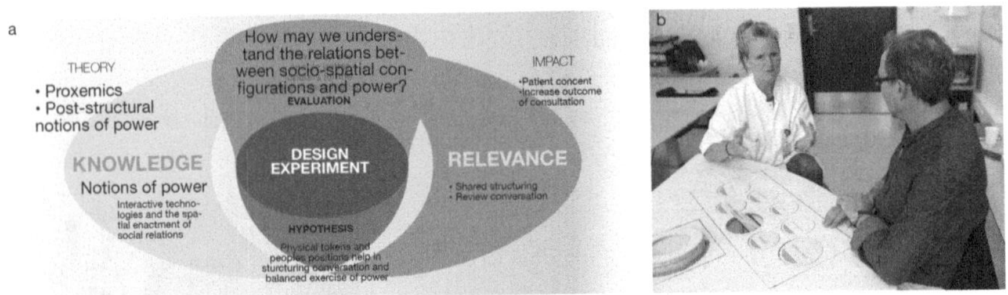

Fig. 4.6 The final prototype and how it relates to the Knowledge-Relevance model

Patient-doctor consultations in oncology were chosen as the domain of design. Thomsen and Schnedler assumed that they would be designing technologies for the doctor to better engage the vulnerable patient in the consultation to obtain informed consent on treatment (Fig. 4.4a).

A range of design activities were initiated including observations Fig. 4.4b, workshops, and provotyping Fig. 4.5b (as Preben Mogensen 1992 called prototyping that aims at provoking discussion) involving relatives, former patients and healthcare professionals. The key design concern of the project drifted in three ways:

- As the observations and workshops revealed that the patients needed better access to the material used during consultation because few patients were capable of remembering what was said during consultations. The varying capacities among patients (some very weak) also pointed to the need to go beyond abstract computer interfaces. The hypothesis of using physical tokens emerged to structure the consultations and balance the exercise of power (Fig. 4.5a).
- The project resulted in an interactive table fitted a microphone to record patient conversations, to serve recollection outside the consultations room and a series of physical tokens (Fig. 4.6b) that helps even very weak patient to participate actively in a consultation. This was the second major drift in the research pursuit pointing.
- The studies also saw a need to complement the design process with postmodern theories of power (Fig. 4.6a). This became the third major drift of the study.

4.3 Design Hypothesis in the Four Epistemic Traditions

There is a rich discussion about the nature of the prototype in constructive design research. We have already distinguished some perspectives. They range from the idea that design objects are physical hypotheses (Stappers 2007; Hummels 2000; Overbeeke et al. 2006) to the idea that they are things to be followed (Koskinen et al. 2011) or props that lead to debate (Jacob Beaver et al. 2009). More recent views have clarified their relationship to research (Koskinen and Frens 2017), to programs (Stappers et al. 2014), and to attempts to push design results from prototypes into products (Will Odom et al. 2016). Again, methodology and those theoretical beliefs that underlie it largely determine the way in which things are done. In caricature, writers from the Netherlands tend to build their view on the issue on their experience in engineering and the sciences, Nordic and American writers on the social sciences, some British authors on art, and in the most recent development, Vancouver-based researchers on post-phenomenological philosophy (Sabrina Hauser et al. 2018).

The debate about the nature of the products of design may be rich, but there has been relatively little discussion about how hypotheses are created. The procedures that lead to hypotheses are, for better or worse, implied in the debate about what manifests a prototype and the methodological discussion. A good starting point is the following observations: there is a range of ways in which designs objects become hypotheses, and these somehow depend on the epistemic tradition presented in Chap. 3, as we will go through in the following.

4.3.1 *Hypothesis in the Experiential Tradition*

The *experiential* epistemic tradition locates the development of a hypothesis in design practice. In this view, design interests come into research through design references that are treated the same way as design references would be treated in industry. Knowledge emerges from the intermediary artefacts that design produce. They are precedents that are carefully curated in the process of creating a design object. Research gives the overall frame and direction to research, but the hypothesis is primarily shaped by practice. The rationale is best expressed in the credo of *Presence project*. If we take Gaver's concept of aesthetic accountability (see Chap. 1) to its logical conclusion, design research is reduced to design referents. The logic is similar to that in design firms. When a design firm gets a commission, market competitors are scanned to see what already exists, and then the design firm curates a collection of possibly interesting objects at point of reference in the process. It breaks these references down into details whenever possible, and then starts to add elements from a variety of sources to the emerging design.

Some tools for working in this tradition have been captured in the concept of annotated portfolios by John Bowers (2012). These portfolios consist of a set or references curated around the emerging design object, annotated with remarks, text, and links to other designs. Annotated portfolios are collections of examples, references, notes, sketches and mock-ups that build up towards a design artefact. Similar to folders design consultants do for their projects (and scientists log books), they provide a practical way to manage design development in research. Their purpose is to keep the emerging design proposal together.

Through annotated portfolios, a hypothesis is grounded in practice of design rather than a process or discourse. In contrast to design practice, annotated portfolios have a research component, however. The portfolios may be maintained from one project to another, and the research program can grow from connections between the design objects and projects. This is a luxury of the research world, where funding is more stable and where researchers can control their problems to a degree that would be difficult in business. Also, the outcomes are research contributions rather than marketable products: they are not tested rigorously, they are at best small series production (and more likely unique), and they get their value from the research story that surrounds them.

4.3.2 Hypothesis in Methodic Epistemic Tradition

The methodic epistemic tradition is exemplified by Pieter Jan Stappers and Caroline Hummels. They see design objects like prototypes as experimental vehicles that are essential to design research. As Stappers (2007) has noted, concepts fail to capture several crucial aspects of design. Only when a design is constructed, it can be assessed properly. There are many decisions that go into turning a concept real, and without a design object — or at least a prototype — researchers can hide behind the concept. Stappers mostly talks about physical products, but the same logic extends to other types of designs like spaces, clothing, interactive systems and services.

While the reason for constructing a design object may be clear, it is less clear what happens in research for it. If we follow the Dutch authors, one of the starting points of any research is literature review, but, as Josephine Thomsen and Jacob Schnedler (2017) show, its scope is modified and transformed by decisions during the design process. These decisions make design research different from, say, psychology, which proceeds by describing and explaining things and proves its point through causalities rather than by building things. If we follow Hummels and other Dutch authors, we get a cue about how literature is processed into a design. In their view, literature is processed into a causal model that leads to hypotheses that can be tested when the design object is evaluated. The prototype becomes an embodiment of a model that specifies it in terms of dependent, independent and intervening variables. In building this model, research grows out of a literature review into a theoretical model that tells what is important for the emerging design, what is not important, and how the variables in the simplified model are related.

In other words, theoretical work creates a structure of meaning that gives researchers design directions, and also a way to observe what happens in testing the final design object, which becomes the *methodic* grounds of the research. Many other things happen around a design object. For example, it can also be regarded as a temporary conversational vehicle. The crux of the matter, however, is that a design object in research is more than the result of studio work. The model originating from a theory concern creates a detailed structure of meaning around the object, and this structure of meaning provides guidelines for interpreting it; increasing predictability and coherence between design intentions and impact of the designed object is ideal of contributions.

4.3.3 Hypotheses in Programmatic Epistemic Tradition

The *programmatic* epistemic tradition situates hypothesis to a research community. The most important proponents of this view have been John Zimmerman and Jodi Forlizzi of Carnegie Mellon University. In their view, the specificity of design research lies in discourse that may have theoretical components, but is also carried by designs and frameworks that extract knowledge from the designs. If they are correct, prototypes and other knowledge vehicles get meaning from research programs around them — thus the name of the epistemic tradition. Any design object can contribute to knowledge. It can target a gap in literature; it can be a critique of the received view; it can elaborate the received view; or it can clarify it by showing how previous research has worked with confounded concepts that need clarification. Theories and frameworks are needed, but so are design objects that serve an essential function in pushing research programs forward.

In this view, prototypes and other 'design things' (Binder et al. 2011) carry not only individual elements of research. They also carry subcommunities of disciplines and in this sense, have a collective dimension. They become key vehicles through which the field of design defines its world. The intersubjectivity of a community that we pointed to in Chap. 2.

This has been the case in several well-known design objects like *UTOPIA project's* participatory methods (Ehn 1988), and *Nutrire Milano's* community prototypes (Anna Meroni and Danila Sangiorgi 2011). These designs get their meaning from theoretical frameworks that discuss the meanings of these designs, and serve as handles for other researchers, but that cannot be understood without the concrete experiences of the design objects. The significance of the cardboard computers of UTOPIA are impossible to understand correctly without understanding the discourse of participatory design of the seventies.

If Zimmerman and Forlizzi get it right, hypothesis in constructive design research build on knowledge shared in conferences, journals, and books. Knowledge sharing clarifies arguments and spreads ideas and practices. This view differs radically from a craft-based idea of design in which design develops slowly through improvements in practice. The view may also explain how design research communities develop their specificities by developing a hard core of ideas that create identities to subcommunities and are non-negotiable (see Koskinen et al. 2011: 39).

Redström (2017) has recently taken this idea further by claiming that design is driven by 'transitional theories.' His view follows the idea that the field of design is like science, characterized on one hand by long periods of stable improvement in the margins, and on the other, abrupt change when there are enough explorations that question status quo. For him, theories are temporary devices in these transitions.

His views are still being evaluated in debate, but Zimmerman and Forlizzi's view is echoed in an old publication that sought explanations for why Milan is the center of the design world. One possible explanation is that in Milan, design is seen as an intellectual discipline in that town. In contrast to, say, Scandinavia, where design is a craft-based activity, in Milan theoretical (usually semiotic) conversation keeps disseminating new opportunities in technology, materials, imagination, history, or poetry into the fore. Creativity in Italian design is a systemic property if this view is correct (see Bertola 2009).

4.3.4 Hypothesis Emerging in Dialectic

The *dialectic epistemic tradition* offers yet another way to locate the origin of a hypothesis. In it, a hypothesis is a result of a dialogue between a multitude of agents. When the object of research is seen as participation rather than strict methodic knowledge production, the hypothesis is seen as an emergent phenomenon designers cannot fully control because it develops in participatory encounters. It is different from the results of a methodic process, theoretical discourse, or an annotated portfolio.

Most authors describe their approach as dialogical. This is how Tuuli Mattelmäki (2006) described her take on probes, and how Ezio Manzini (2016) has described Politecnico di Milano's approaches to service design and sustainability. For constructive design researchers in this tradition, a hypothesis is a construct that is built together with people involved in the design process. Designers have only partial control over the hypothesis, and this departure from control is made in name of participation and democracy — dialectically pointing to a potential future. Another word that has been used is debate, as in *Material Beliefs* (Beaver et al. 2009) and its Design for Debate framework, which implies a form of discussion in which several parties come together to talk about issues. They start from different — and sometimes even adversary — positions. In the course of the debate, they get familiar with the viewpoints of other parties, which may lead to consensus that may initiate change. In the language of this chapter, this consensus and the initiative to change are elements that form the hypothesis.

It is the adversarial aspect of experiences that has gained attention after Carl DiSalvo's *Adversarial Design* (2012), which painted a picture of design as a process that brings adversaries to the same table. It lays bare their differences, and after confrontation, leads to a program that hopefully brings about change that lets the participants better achieve their collective ends. The same political ideal has animated most writing about design activism, which finds its roots in the disruptive aesthetics of Rancière (2004), who in turn builds on Louis Althusser's structuralist Marxism. At bottom, these approaches are participatory and aesthetic rather than political. However, exactly how change is to be accomplished remains by and large open.

The one exception to this generalization is Scandinavian participatory design of the seventies and the eighties. It clearly aligned with Marxist and Social Democrat politics, which is well illustrated with one of its foundational projects UTOPIA (see Ehn 1988). The project worked with graphic workers' unions. Graphic work was going through a technological upheaval to computer-based technologies, and UTOPIA aligned with trade unions in an attempt to salvage the skilled factory jobs new technology was about to destroy. The political aim of the project was to combat deskilling and indirectly maintain the negotiation power of the unions against capitalists. There was no need for a theory of change; it was implied in the political commitments of the researchers.

The project's long tail in design research builds on the way it discovered the wisdom of using mock-ups and other 'design things' (Binder et al. 2011) that created a language game both graphic workers and designers could understand. If we look at its design proposals in terms of a hypothesizing, the work centered around mutual learning that served as basis for collective hypothesizing — the *dialectics* that build the common contribution. In these dialogues, design acts came to be a joint articulation of interests of a significant, well-paid group of workers whose work was about to be deskilled in the interests of owners of corporations like newspapers and printing houses. The hypotheses of the project reflected the dialogue in the workplaces, but also larger-scale social concerns.

Designers became activists who did not hide their politics into technical expertise or aesthetic theories but saw participation as a crucial element in encouraging hypothesis to emerge from mutual learning. A hypothesis like this is temporary and specific to the involved community, and it is not usually meant to be defendable in universal social terms. Knowledge gets measured in terms of whether it works through participation rather than whether it is valid in the traditional sense of the term: an expression that cannot be contested because it approximates truth. Design artefacts, in their part, carry knowledge in these discussions, but are not meant to serve as testbeds, or as carriers of a research community. Rather, they gain a meaning from the community whose interests researchers want to serve. In this tradition, research constructs like hypotheses get their meaning in communities of practice rather than theory alone.

4.4 Hypothesis in the Knowledge-Relevance Model

The purpose of the K-R model is to facilitate a continuous process of framing and reframing the actual research work. It center-stages the critical 'conversation with the material' (Schön 1983) when constructive design research oscillate between, hypothesizing, experimenting and evaluating nurturing both relevance and knowledge production. The model builds on research and on an alternative model developed a few years back (Bang et al. 2012), and it is well in line with recent discussions. For example, Kees Dorst (2015) proposes that design is about framing and re-framing. In the view, expert designers can concurrently propose and test numerous hypotheses and questions while suspending decisions on solutions in pursuit of the most relevant question. Similarly, in the Knowledge-Relevance model, hypothesizing is seen as an ongoing process that happens in tandem with framing questions. The process is flavored by the researcher's motivation for doing the research, informed by provisional theories, anticipated measures of impact and recurrently challenged by the sobering question of 'what question is this experiment an answer to?' — the things you hold yourself accountable to.

We see the model as a map that helps researchers to clarify their contribution. It serves as a tool in the research process aiding the researcher to understand what is contributed (relevance and knowledge), what are the driving forces of the discussions and claims brought forward (hypothesis and evaluation), and what measures apply (impact and theory). And finally, it serves as a good reminder of what to take into account when designing a research project.

Looking at the model through the lens of drifting while construing a hypothesis, we can note that it only forms the ground of drifting, but it does not determine how drifting happens. The way in which designers define where they drift depends on the epistemic tradition that defines their concept of knowledge. For example, some authors — most recently, Johan Redström (2017) — have stressed design as a research discipline that gains its force primarily from theoretical statements. Others — like Gaver — contest this and locate the source of the force primarily in design practice instead. Participatory researchers see the hypothesis as a property of the community they work with. Most authors stand between these extremes and if we read carefully, this is the case of authors like Redström and Gaver too. Their arguments are sharp, but their practice is more complicated than these arguments.

Again, these differences point towards a larger context of research. Constructive design research sometimes happens in purely academic projects. Equally often, it happens in tandem with research projects built together with industry (or trade unions as in *UTOPIA Project* described by Ehn 1988). Design also contributes to cultural and even artistic production outside the market. The K-R model allows for the co-existence of diverse concerns and points to design experiments as the propelling factor of constructive design research; the key subject of the following chapter.

Chapter 5
Ways of Drifting in Design Experiments

In Chap. 4, we presented a Knowledge-Relevance model (K-R) that maps design activities in terms of evaluation and hypothesis construction, and knowledge and relevance interests. We analyzed some of the ways in which hypotheses construction takes shape in the four traditions we have identified in Chap. 3. In this chapter our focus will be at the very heart of the model: how design experiments articulate research interests, how drifting happens in experimentation, and how drifting happens between design experiments. Based on the corpus of PhD dissertations that form the foundations of this book, we provide a typology comprised of five types of design experiments. We will label these as *accumulative*, *comparative*, *serial*, *expansive* and *probing*.[1]

Our typology accounts for relations between and across major cases, iterations and immediate thoughts embodied in sketches and prototypes. The purpose of the typology is to provide an overview that respects and accounts for the characteristic ways in which constructive design research happens: process-loops where hypothesis, experiments, and insights concurrently affect one another. We argue that when knowledge from experimentation is continuously fed back into the research, it drifts, but through a pattern captured by our typology. The argument against formal strict methodologies is an often-encountered theme throughout the dissertations that we draw upon. Most writers argue that design research is more than problem solving because design challenges and problems tend to change significantly during the design process.

[1] This work is based on the five ways of drifting' paper (Krogh et al. 2015) and provides a general outline of the characteristics which point to the methodological roles that design experiments and design work may contribute in constructive design research.

P. G. Krogh, I. Koskinen, *Drifting by Intention*, Design Research Foundations,
https://doi.org/10.1007/978-3-030-37896-7_5

5.1 Drifting in Constructive Design Research

In classical scientific methodology, researchers are encouraged to create interesting and novel hypotheses, but use well-known methods in pursuing knowledge. 'Drifting' is seen as a flaw because it introduces method variance that makes the results of the study unreliable. Redström (2011) also points to this. In design, however, 'drifting' is usually seen as a positive quality measure. It tells the story of a designer capable of continuous learning from findings. This, in turn, signals sensitivity to adjusting conditions and a good 'nose' for hunches. Likewise, constructive design research that does not drift misses this general hallmark of recognized professional excellence, and may fail to gain respect from fellow designers. Yet, tolerating 'drifting' raises another built-in dilemma between whose criteria leads research, designers' or researchers'. For the latter, the question is the degree in which researchers can trust the results of constructive design research.

When examining the existing research literature, one is left almost empty handed when searching for detailed accounts on processes and basic constituents of design experiments. As noted by Redström (2017) they take up many forms. Zimmerman and Forlizzi (2008) attempt to develop a formal account of methods used in constructive design research. They suggest that a foundational distinction should be made between two different methodological approaches:

i. A philosophical approach, where researchers wish to 'investigate a previously articulated theory through a process of making' (e.g. 'ludic interaction', 'rich interaction', 'aesthetics of interaction', etc.); and
ii. A grounded approach, where researchers focus 'on real-world problems by making things that force a concrete framing of the problem.'

This meta-level classification tells how practitioners see experiments. It remains abstract, however, and provides few insights into how design practice and experiments are conducted in either of these approaches.

In order to align the methodological foundation of constructive design research with the practices of professional design Brandt and Binder (2007) and Redström (2017) suggest that design experiments in design research can be better understood as being framed by a 'program' and 'research question.' They assume that research-through-design (which we here term constructive design research) can be modelled in the same way as a design project where a program or brief is typically used by a client to formulate an assignment for a professional designer. While they put much effort into defining the notion of the program and research question, they say little about the design experiments themselves. According to them *experiments* in constructive design research are examinations of *questions* residing in *research programs*. Bang et al. (2012) point to the act of hypothesizing as a fruitful and 'direction-providing' activity in design. Whereas the work of Zimmerman and Forlizzi (2008) articulates design research as being theoretically, technically or empirically inspired, Bang et al. (2012), Bang and Eriksen (2014) follows the line of Brandt and Binder, and Zimmerman et al. (2010).

Along the taxonomy of 'Question, Program and Experiment' promoted by Brandt (2006), Brandt and Binder (2007) and Redström (2017), Bang and Eriksen (2014) point to the changing characteristics of experiments as they are conducted at different times during a constructive design research process. In their account experiments may serve three different purposes, including being framing vehicles of the research; being testbeds of the limits of the research interest/ program that may cause the research to drift; and being exemplars of knowledge production. While Bang and Eriksen point to the varying roles and characteristics of experiments in constructive design research, they give little guidance to how experiments interrelate and how experiments are developed over time.

In line with Bowers and Gaver (2012) we believe it can be of great importance to declare how one got there — how the design project drifted and gained insights unintended by its original pursuit — and the knowledge one developed in doing so. In line with Gaver (2012) and Koskinen et al. (2011), our point of departure is that knowledge production in constructive design research is characterized as fallibalistic.[2] For example, any sketch can be seen as a question examining provisional hypotheses in the manner of Piercean 'abductive reasoning' (1979).[3]

5.2 Drifting Through Design Experiments: Five Methods

The heart of the K-R model in Chap. 4 is design experiments. We argued that the model enables researchers to continuously map and re-map their research activities as a conversation between hypothesis construction, experimentation and evaluation, assessed in relation to both knowledge theory and impact, in the ambition of being relevant and producing knowledge. Here, we build on this model and provide a typology of design experimentation. The typology is originally derived from the analysis of 10 PhD dissertations (Christian Dindler 2010; Ott von Busch 2008; Kristina Niedderer 2004; Linda Worbin 2010; Ambra Trotto 2011; Philip Ross 2008; Majken Fogtmann 2011; Joep Frens 2006; Aviaja Lynggaard 2012; Anne Louise Bang 2011) and in this book complemented by the corpus of dissertations included in Chap. 10.

[2] We use the term in line with Richard Feldman (2003) that knowledge do not require certainty and that claims and theories are to be considered provisional and open to change and revision.

[3] Here we bridge the design practice of sketching with C.S. Pierce' notion of abduction (1979). In our terms a sketch is the physical manifestation in any material of a design concern, which enters into conversational dialogue with the designer as suggested by Schön (1983). Sketching in this way is neither inductive or deductive, by may be considered abductive in the sense that it is a bold suggestion of a likely structure/ pattern at hand. Dorst (2015) suggest the notion of 'designerly abduction' pointing to the designers concern of making suggestions for a potential future; bridging the gap of logic reason and creative proposals. In Chap. 7 we get back to the various notions of bridging reason and creative suggestive practice in the context of constructive design research.

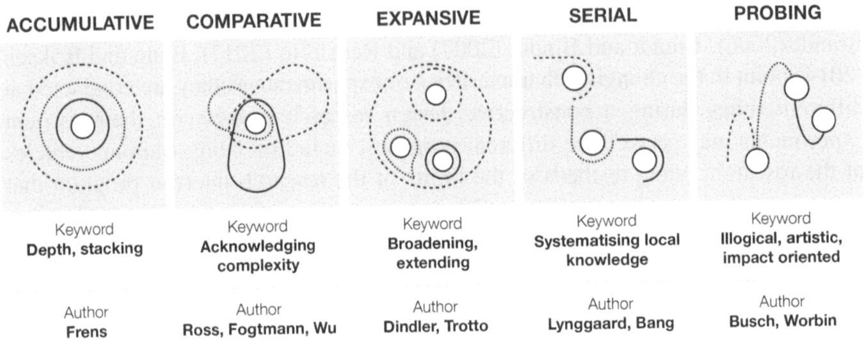

Fig. 5.1 Ways of drifting

As we have pointed out in the introduction of this chapter, the typology of drifting describes five distinct methods of knowledge production through design experiment: *Accumulative*, *Comparative*, *Expansive, Serial* and *Probing*. Fig. 5.1 includes a graphical model of each of the methods accompanied by characteristic keywords and author names of PhD theses that exemplify the methodology. All the presented methods call for 'drifting,' but to a varying degree. The first category 'accumulative' is the least forgiving and 'probing' allows for the largest degree of 'drifting.'

5.2.1 Accumulative

The accumulative method of experimentation is particularly well illustrated by the work of Frens (2006). His design experiments study how tangible interaction might enrich the experience of using a camera (see Fig. 3.3).

The design sketches and models are focused on testing specific parts of the camera and the rich interaction framework. His work is carried out in a laboratory setting in which he evaluated his camera variations to learn about their impact on the users' cognitive qualities rather than for contextual appropriateness. The work shares many similarities with what happens in technical lab settings where one particular variable is studied, and potentially disturbing elements are excluded for the sake of clarity and rigor in the study. What the study loses in relevance it gains in depth. The increasing depth of knowing derived from every experiment is iteratively build (layered, stacked) into the next generation of the same version of the camera. A way of stacking knowledge where the final artefact embodies the total knowledge accumulated through the constructive design research process.

5.2.2 *Comparative*

The method of comparative experimentation is an often-deployed merhod and can be found in the work of Ross (2008); Fogtmann (2011); and Wu (2017). They all explore their subject by means of a number of design cases — working from or towards a shared platform of comparison.

Whereas Ross is interested in ethical and aesthetic aspects of interactive products (in particular lamps), Fogtmann describes the concept Kinaestetic Empathy Interaction (KEI) through a series of design cases each highlighting distinct and overlapping qualities of KEI (Figs. 5.2a, b). Wu wanted to transgress a paradigm that sees services as products to be delivered, and proposes to see services in discursive and provocative terms. When drifting happens in the comparative method it is mostly relate to the shaping of cases. The strength of the comparative method is that it makes it possible to cover areas and aspects not yet dealt with in previous experiments and to expand knowledge gained in them.

Each design experiment should reveal as-of-yet undocumented additional qualities of a concept and confirm some previously found qualities. In totality, the comparative experiments ideally describe a novel concept, qualify phenomena, unseen quality or add a theoretical distinction to known theory. A good keyword to characterize this model might be 'acknowledging complexity,' which expresses the idea that the design experiments explore a concept by embedding it in many different environments. Furthermore, the method reveals that a lot of experimental design work done to shape a case — the characteristic of drifting in the methodology – will not necessarily find its way into knowledge production.

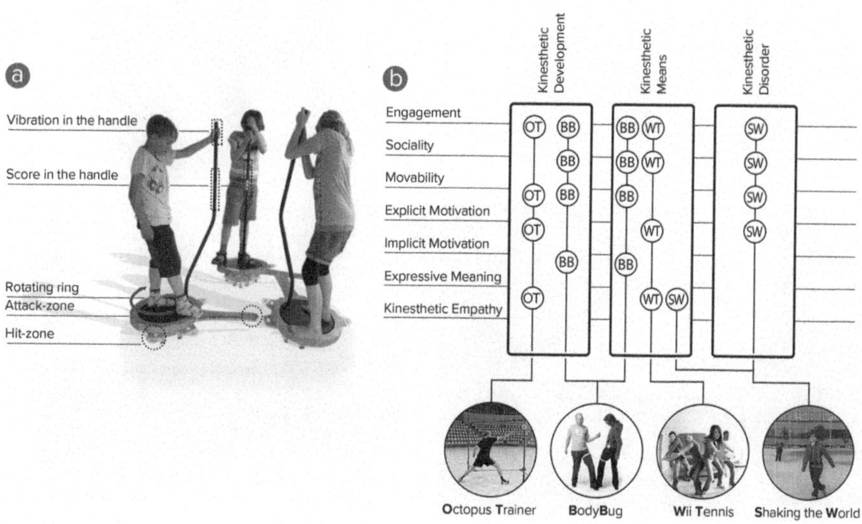

Fig. 5.2 Fogtmann's framework of KEI

5.2.3 Expansive

This method articulates the identification of an area as-yet-uncovered with the ambition to reveal its qualities. This mode of investigation resembles the work of nineteenth Century geographers and biologists mapping unknown areas. The work of Christian Dindler and Majken Kirkegaard Rasmussen (2015) serve as good illustrations. Unlike in accumulative experimentation, there are no strict successive or linear research designs to follow. Rather, a more complete picture emerges as mapping goes on and new data comes in.

The keywords we like to use to characterize this method are 'broadening' and 'extending.' Rather than deepening our knowledge of a domain, the method widens our perspective and extends the concerns designers should include in their praxis. For example, Majken Kirkegaard Rasmussen (2015) explores concerns and experiential qualities of what she uniformly terms 'Shape Changing Interfaces' (Fig. 5.3) through four design studies including initial form exercises, mock-ups, 'wizard of Oz' concepts and prototypes of interactive actuated shape-changing products. Christian Dindler expands our idea of what 'engagement' might mean in interaction design through three diverse experiments. For these experiments, he designed engaging artefacts and developed participatory design methods to understand engagement.

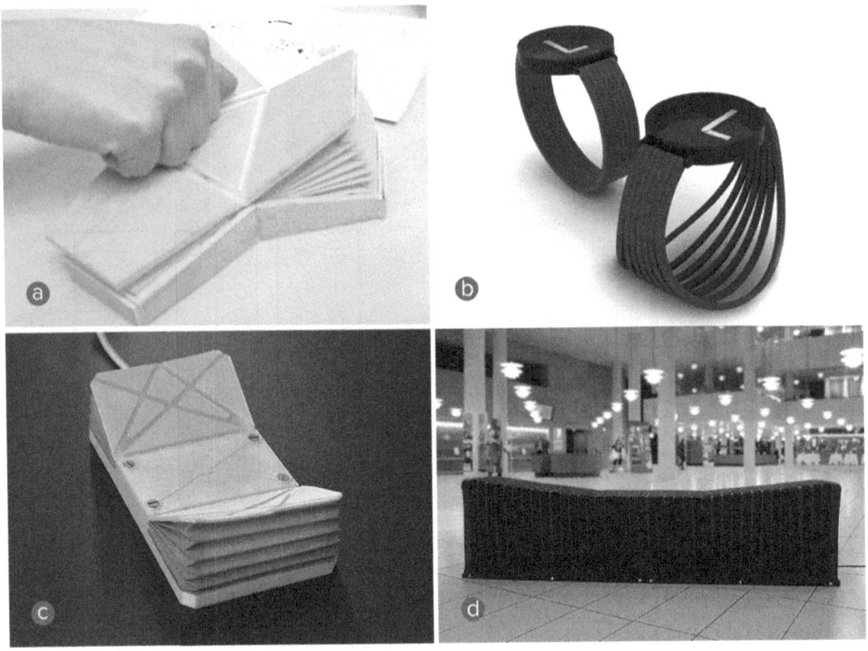

Fig. 5.3 Top left: coMotion, top right: ReFlex, bottom left: KnockKnock, bottom right: CAREss

5.2.4 Serial

The method of serial design experimentation builds on a practice designers are well aware of, seriality. In serial experimentation, design objects as well as contextual variables evolve over the course of the study. The main difference to accumulative and comparative experimentation is that in those two forms of experimentation, researchers want to vary contextual variables, but leave the basic design object as stable as possible. The aim is to study how it changes as a function of changes in contextual variables. In serial evaluation, this possibility is ruled out. The method is more akin to product development than science; in the former, each generation of products typically introduces many improvements.

In serial method, knowledge generated in a design experiment is integrated in the following experiment. In the work of Aviaja Lynggaard (2012), for example, we see how each successive experiment builds upon its predecessor. Each study generated insights that pushed the work forward. These pointers provided large and small contributions to the overarching interest in 'homing tactics.' More specifically, on the basis of a series of ethnomethodological studies, she identified a set of tactics for 'making home' — or 'homing' as she called it. Rather than following a strategic approach, she adopted an opportunistic and pragmatic approach where the results from each experiment were explored in successive design experiments. These were based on pragmatic concerns (time, technical request, budget, company interests etc.) but also on the experiments' capacity to yield additional contributions to the overall research interest.

Likewise in the work of Anne Louise Bang (2011), we find a series of experiments in which each experiment continually builds on the previous one. She was mainly interested in exploring the emotional value of applied textiles from three perspectives: 'textiles as material'; 'textiles as part of an object'; and 'textiles as part of an object in an environment.' She invited a variety of stakeholders to participate in her exploration. Her objective was twofold. One objective was to develop an in-depth knowledge of emotional aspects of textile design. Another objective was to build stepping stones helpful in developing a structured approach to inviting stakeholders to participate in industrial textile design processes (see also Bang and Eriksen 2014). A key characteristic of this method is 'systematizing local knowledge.'

5.2.5 Probing

Exploiting opportunities and exploring design ideas as they emerge through design work is also what characterizes the final method described here: probing. This method is even more radical than the serial method. In serial experimentation, there was a path from one study to the next, even though both the design piece and the environmental conditions could change. In probing, there is no need for even this

Fig. 5.4 (**a**) Worbin experiment on thermochromic textile dye, (**b**) von Busch DYI fashion workshop

amount of consistency; the method is more akin to the way design firms work. They cannot predict and expect much consistency from one project to another. Probing accepts this fact; in it, both the design and the environment can change from one study to another. A research world equivalent is *a fortiori*, the method of a case that through similarity borrow the argument of a stronger case. The success of this method depends on whether the designer can find a red thread that binds all the different arguments together.

The approach is widely used in arts and crafts oriented design research and it is well documented in Linda Worbin (2010) and Otto von Busch (2008). Yet, it is only when we examine probing in relation to the other four methods that its methodological value for design research can be fully grasped. What often characterizes this methodological approach is a personal motivation and engagement in the research pursuit, where the research activities are points of impact in a research field larger than what a single research project can be expected to cover. The choice of experiments by Busch (2008) and Worbin (2010) can be characterized as 'illogical', 'artistic' and 'impact oriented.' (Fig. 5.4).

Worbin is interested in the merging of IT and textiles in the very broad sense. Through a number of experiments she highlights recurring and important aspects of this mélange of material and experiential properties and qualities. Busch is interested in hacktivism as part of democratizing fashion production. Both the theses and, presumably, their doctoral studies are logically structured endeavours exploring the qualities of a field. However, these two are characterized by selecting in an almost eclectic manner wicked, ir-reductive and self-contradictive design settings derived from pursuing opportunities in the environment (as a professional designer would do). From both a practice and research point of view this strongly test their subjects. On the basis of such experiments they make contributions valuable to design research and foster curiosity for the field itself and its neighboring areas.

The method is typical to another context too. In dissertations done with industry, researchers may not expect full control over their research design. These dissertations

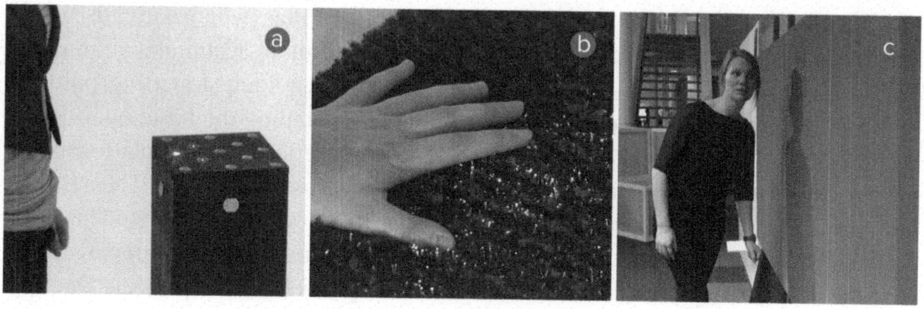

Fig. 5.5 Eva Deckers's PhD thesis: (**a**) PeP, (**b**) PeR, (**c**) PiA

typically become jump from one study to another, and the creative task of the researcher is to find a storyline that binds them together. Good examples are Katja Battarbee's by now classic dissertation *Co-Experience* (2004), which built on a miscellaneous set of studies that finally came together when she discovered the concept of co-experience, which became the thread that gave coherence to her case studies. Another example is Simo Säde's *Cardboard Mock-Ups and Conversations* (2001), which told the story of user-centered method development for a design firm. The methods were developed in client projects with a company.

5.3 How the Methods Work Over a Design Process

So far, we have talked about drifting as is it consisted of a set of distinct methods. However, many researchers have used several methods in their work. In the following we will show this by analyzing two cases: *Perceptive Qualities in Systems of Interactive Products* by Eva Deckers (2013) and a short master thesis project carried out in Denmark during 2017 by Thomsen and Schnedler (2017). The latter was also used as case in Chap. 4 to illustrate the K-R model.

5.3.1 Comparative — An Example

Deckers (2013) was interested in 'perceptual crossing,' a phenomenological concept she wanted to clarify and evaluate. She was in particular interested in how designed artefacts could be perceived as being perceptually aware of the presence and the activities of human beings. Her work is a good example of how design research can be grounded 'empirically' and 'philosophically' (Zimmerman and Forlizzi 2008). This continued interaction of perspectives and activities is also what Deckers (2013) term 'Leitmotiv' in her dissertation (Fig. 5.5).

The first two cases of her study were the Perceptual Pillar (which she calls 'PeP') and the Perception Rug ('PeR'). Both were developed in an accumulative manner. She was confident and well aware about the phenomenological starting point of both studies, and directed her efforts to refining and detailing the basic qualities of the artefact. The design of both PeP and PeR can be considered as platforms for theoretical explorations with regard to the concept of Cross-Perception. The description of PeR is complemented by Surfacing Sound, a closely affiliated student project called PiA (an intelligent audio system), which adapted the concept of perceptual crossing to listening music at home. By including this student project, Deckers includes an auditory experience into the otherwise visual argument about perception.

Throughout the dissertation Deckers talks about her activities in the dissertation as 'designing.' Indeed, she reports her design outcomes, but also those theories that informed her analysis. The models and conceptions are intended to inform designers in the future. From a methodological point, the inclusion of the PiA student project gave her method an 'expansive' form. From the perspective of Decker's argument, Surfacing Sound also showed that her framework could be used by third persons. She could argue that models and conceptions are indeed useful for designing cross-perceptual qualities into products because of the student project.

5.3.2 Accumulative — An Example

Based on the master thesis of Thomsen and Schnedler (2017), we point to how the adoption of the drifting methodology perspective helps declare the character of constructive design research work happening also at a detailed level. Whereas the work of Deckers (2013) covers larger ground and reports on years of research and includes the reporting on several design artefacts, Thomsen and Schnedler's contribution was based on the development of a single case (Fig. 5.6).

The thesis set out to explore complementing theoretical aspects to socio-spatial qualities identified by Krogh et al. (2017). Thomsen and Schnedler did this by providing interactive means for doctor-patient consultations in oncology. The design result of the thesis work was an interactive table facilitating audio recording of the consultations tagged by the use of physical tokens. The tokens furthermore served as means for structuring the consultation. Through a detailed account on activities Thomsen and Schnedler were able to point to a drift in their research question, and analyze how and what activities caused the drift.

The methods that informed the drift were observations of consultations; interviews; workshops with patients, relatives, and healthcare professionals; literature studies; and provotypes. More than 25 observational sessions and interviews were conducted 'comparatively' in dialogue with literature studies. These studies showed that the study had to explore 'the exercise of power' to understand the configuration of healthcare personnel, patients and relatives during consultations. This framing was explored in an 'expansive' manner through initial prototype sketches in order to understand the specific character of the situation. Key qualities were then

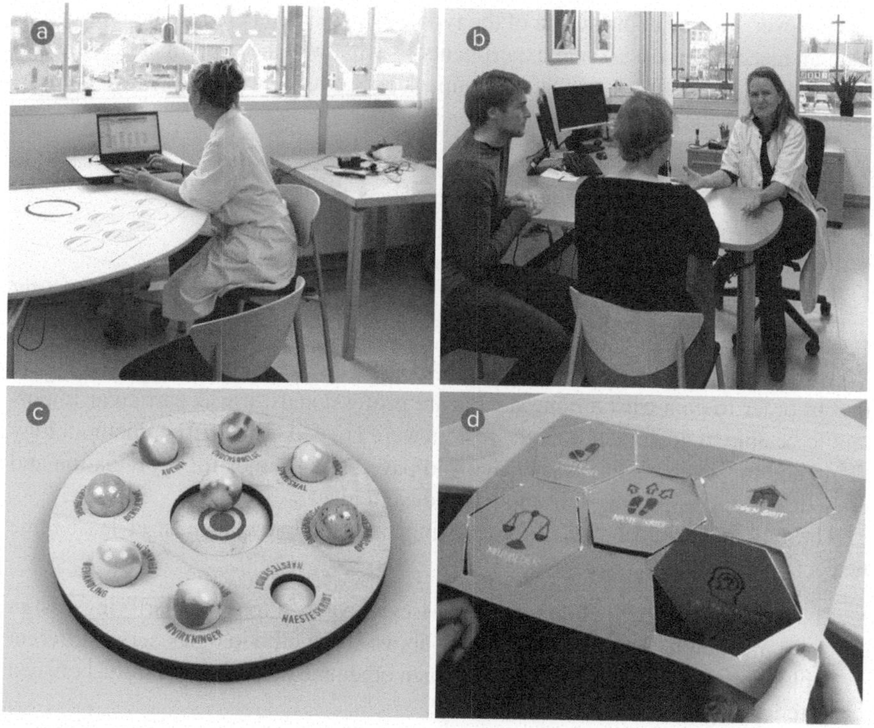

Fig. 5.6 (**a**) Interactive Consultation Table, (**b**) initial observations, (**c**) provotype, (**d**) mock-up of table tokens

deepened through 'serial' sketching producing a set of provotypes (Preben Mogensen 1992) that were explored in an expansive manner in a variety of meetings with grouped stakeholders. It was this methodologically well-documented process that enabled Thomsen and Schnedler to transparently argue for the drift from asking how can we use IT to support doctors in the orchestration of a consultation to how can an interactive interior comprising connected intelligent surfaces and objects facilitate a more equal relationship between doctor and patient?

The process of sketching and developing the actual artefact oscillated between 'serial' associative flows of idea generation. It also integrated 'comparative' discussions. As the design concept became more and more clear, well-argued aspects were developed *accumulatively* in order to gain further depth in argument.

5.4 What the Typology of Drifting Methods Reveals

This chapter has described five methods of drifting. We believe that this typology will help constructive design researchers to understand better their research processes. The reason for adopting a typological perspective was that we wanted to articulate how research is pursued.

However, the key contribution of the typology presented here is to help researchers plan research processes and experiments — what will the experiments bring and how to plan for certain experiments to unpack concerns? The typology also helps to define whether the researcher wants to:

- Increase depth (accumulative);
- Stress collective and individual characteristics (comparative);
- To learn from experiments to thicken a concept (serial);
- Broaden the scope of the study (expansive);
- To discover and unpack a research interest and how one may imagine connecting diverse experiments when the topic is ill-defined (probing).

Thomsen and Schnedler's study is relevant for us in another way too. It shows that in order to gain trust among healthcare professionals, it was particular important to be able to document how the results were created. The drifting methodology facilitated this exchange of knowledge and paved the grounds for appreciation and open discussion of proposed framing and design ideas.

The typology contributes to the existing body of knowledge in at least three respects:

- The typology of methodological processes serve as a tool for clarifying design experimentation. As pointed to in the above the methodologies serve different exploratory pursuits that complement each other and helps declare shifts between different experimental *modus operandi*.
- It allows for a concise description of different knowledge outcomes that may result from design experimentation: depth or stacking of knowledge; acknowledging complexity; extending knowledge of a certain area, and so on. Typically, in the research literature we have consulted such descriptions are not given. Rather, knowledge outcomes are classified generally in terms of for instance, 'nascent theory,' 'conceptual frameworks,' 'guiding philosophies,' or 'design implications' Zimmerman and Forlizzi (2008).
- Beyond providing a means to distinguish methods of experimentation in design research, the above typology also depicts a spectrum of methods that have a strong or light foothold in the wider world of science and technology. Accumulative, comparative and expansive experimentation have a strong foothold within technological research, which usually builds on the idea that knowledge comes from controlled experiments. Serial and comparative experimentation are more typical to the social sciences.

In many ways, we see design experiments as turns in conversation. They start form a base in literature, run through iterations that take researchers from knowledge to society and back, and finally end up adding something to existing knowledge. Things designers create during this process are boundary objects that help people to find a common language, but they are also more than that. They are heuristic devices that observations made lead during the process often lead researchers to question their theoretical assumptions. Their conceptual outcomes are obviously not models in any formal sense; they do not build on clear axioms. Rather, they are like physical, virtual, spatial and processual descriptions of theory that match the intuitive ideas of the original theory. We believe this is what the logician and philosopher Patrick Suppes (1960: 9) might have said about them in his philosophical treatment of models in empirical science if he had talked about design. In design, we tend to think, this is a permanent state: We find it hard to believe that design, a discipline built on things that exists, one day would be able to find a true axiomatic or even paradigmatic base.

In this chapter we have first described the typology of drifting native to design experimentation. We have then showed how drifting in design, knowledge interests, and research contributions happens at a more detailed level. The ability to describe the ways in which they drift, we think, will help the constructive design researcher to collaborate better with other disciplines. This ability gives them flexibility to adopt different perspectives, see similarities between seemingly diverging ideas, and find ways to create solutions that are acceptable to several stakeholders. Drifting in a sense becomes a useful negotiation tactic when there are divergent and even conflicting perspectives on issues in the studio, product development department, or in research. The ability of designers to articulate their visions in terms of physical and virtual things also gives them ways to bypass ingrained patterns of argumentation. The danger to avoid, we think, is to leave products on their own without a narrative that gives people tools to talk about them. The typology of methods presented here also helps to track and plan the ways in which design experiments happen. It finally contributes a visual way to make sense of the epistemological ground of research practices in design, but also artistically inclined and aesthetic practices.

Chapter 6
Four Epistemic Traditions of Evaluation

On what grounds do we judge whether a theory for design is useful, valuable or successful? What is validity in constructive design research? What is the role of theory produced from design? Chapter 4 dealt with ways of construing hypotheses and how the K-R model may help map the constituent parts of a research process. Chapter 5 presented the typology of ways of drifting explaining how design experiments inform and urges design researchers to drift. This chapter turns to how design researchers can *evaluate* and justify their claims about knowledge. And completes our core trilogy of dialectic activities serving the dual ambition of relevance and knowledge production in constructive design research.

Our claim in this chapter is that evaluation takes different forms depending on the underlying epistemological tradition. If we are correct, evaluation must be conducted in different ways simply because the yardstick of what constitutes a contribution is vastly different in the identified traditions. Drifting, in turn, depends on the concept of knowledge that is held important in these traditions and the trajectory of experimentation.

Looking at literature produces some insights. Clearly, there is no agreement about evaluation. If we look at the corpus, we find diverse views. Some of these are incompatible at bottom. At one end of the spectrum, evaluation is practiced as systematically and rigorous as in controlled lab settings (Frens 2006; Ross 2008), while at the other end research through design is dedicated more to activist reforming of practice and the crafting of manifestos than to evaluation (Trotto 2011; Von Busch 2008). The same can be said about research literature. There are calls for adapting more rigorous methods from science, technology, social science and the humanities (Zimmerman et al. 2010) and there are calls that deconstruct the idea of rigor with philosophical and sometimes artistic references (Gaver 2012).

The good news is that this diversity can also be seen as a sign of maturity. By now, design researchers have developed a set of approaches to evaluation. In this chapter, we want to appreciate the heterogeneity of evaluation practices and criteria that is characteristic of constructive design research. Yet, at the same time, we argue that although constructive design research may be judged against notions of validity

P. G. Krogh, I. Koskinen, *Drifting by Intention*, Design Research Foundations,
https://doi.org/10.1007/978-3-030-37896-7_6

(Jillian Hamilton and Luke Jaaniste 2009; Niedderer and Roworth-Stokes 2007: 5), a more flexible way is to see it in terms of accountability (Koskinen and Krogh 2015). Diversity in evaluation does not preclude a coherent formal account of this diversity. In the following, we provide a brief account on recent debates on evaluation in constructive design research which is complemented by a scheme modelling how traditions and ideals of contributions, methodologies and approaches can be mapped.

6.1 Evaluation in the Experiential Tradition

Artistic and whimsical evaluative practices were called eclectic by Markussen and his colleagues (2015). Here, we talk about probing to be consistent with the terminology introduced in the previous chapter. These practices are akin to showroom methodology (Koskinen et al. 2011). In the *experiential* epistemic tradition, a constructive design researcher cherishes the fact that people form their opinion of a design object differently from hers. She may collect these opinions and use them in design (e.g. Wallace 2007). She may equally well let them go and be happy about the fact that people have a right to form their independent opinions, however. The same principle applies to her own creative practice, which may go undescribed.

Within this epistemic tradition, researchers have created several evaluation practices. Below, we illustrate some of these with three examples. Summatavet (2005) aligns closely with craft world practice, von Busch (2008) with art world practice, and Mäkelä (2003) with phenomenological philosophy.

When probing is the method of evaluation, researchers may refuse to open parts of their creative process altogether. For example, Kart Summatavet's *Folk Tradition and Artistic Inspiration* (2005) had a constructive part, which took the form of three exhibitions in which she explored jewelry in three stages of a woman's life: the childhood, maturity, and old age. Her design objects were exquisitely crafted. They were juried by professional goldsmiths, exhibited publicly in high-end museums in Helsinki and Tallinn, and they were assessed by a researcher, by an artist, and to a lesser extent, by the public. The exhibitions were based on the learnings from her folkloristic fieldwork in Estonia. Summatavet, however, consistently separated her field studies and art work, and refused to show the connection. This was accepted in the design school context in which she was working. If we respect her integrity as a design researcher, we have no right to treat showing the path from inspiration to art works as criteria for success. Instead, the main criteria must be the art works.

Von Busch's dissertation *FASHION-able* (2008) focused on hacktivism and consisted of several studies including an exhibition, a residence in a shoe factory, a failed attempt to redesign a Turkish fashion firm, and establishing a student magazine. These were interpreted with concept like heretics and rhizomatic validity in the line of postmodernism of the eighties. In von Busch's thesis, there is hardly any evaluation in the traditional sense. He is interested in reforming the fashion system from bottom-up and democratizing it by making it possible for the man on the street and professional amateurs (so-called pro-ams) to have a say in the shaping of fashion. One of his central arguments is that to let this happen, a new anarchistic approach has to be invented, which enables collaboration between fashion designers and non-designers and which he refers to as 'hacktivism.' Busch draws upon theories of DIY, software hacking and political theories to formulate the conceptual foundations of hacktivism and he sets up a series of open workshops and exhibitions to explore if people and programs are able to reform the fashion system. The workshops are not tied together by certain logic of progression or expansion, but exist as singular events planned to explore the fundamental question: is it possible to turn fashion into hacktivism?

In the evaluation of one of these workshops held during the Hackers and Couture Heretics Exhibition, Busch concludes that it has been a success for two reasons. First, it was easy for people and bystanders to participate due to the use of simple methods such as re-design, button exchange, shop dropping etc. (ibid., p. 219). Secondly, the workshop proved that 'DIY does not have to look grubby or lack craftsmanship' (p. 222).

When looking at probing as a way of evaluating the above observations tell a great deal about its character. It rests entirely upon subjective judgments, and the purpose is first of all to highlight that people actually participated rather than how they did it. The ultimate goal in Busch's dissertation is to demonstrate that hacktivism is possible and that it does not rule out aesthetic quality. Probing evaluation is thus used by Busch, not to document the validity of results, but as means for articulating a new approach and aesthetics founded upon DIY, activist philosophy and political theory. The notion on probing in evaluation is elaborated upon in the following section.

In the dissertation *Memories on Clay* (2003), studying womanhood through a series of three exhibitions, the ceramicist Maarit Mäkelä working in the experience-based epistemic tradition created a philosophical evaluation method. Initially, she built on feminist art theory, and built an exhibition around public stereotypes of women. In this exhibition, her work consisted of sexualized images of women painted and printed on ceramic objects. After the exhibition which like Summatavet's was juried and evaluated by experts, she went back to the studio and reflected upon the feedback. These reflections led to a second exhibition, which focused on the female body. Her works incorporated materials from her own body, including hair and menstrual blood. The stages were similar to the first exhibition. The third, and final exhibition took place in a small cellar gallery in the outskirts of Helsinki's city center. In this womb-like space, she exhibited a series of works that were inspired by auto-ethnographic research into the maternal line of her family. She turned these studies into fictional objects and called the method auto-fiction. After the third exhibition had been evaluated by experts, she went back to her den and reflected the outcomes of not only the last exhibition, but the whole path from the first to the third. She called her method 'retrospective gaze,' and openly admitted that it was mostly fictional. In this method, it was impossible to break the meanings of her work into components by where they came from, and when they were created. The method was phenomenological and was for certain murky by scientific standards.

Yet, it produced a consistent storyline that cantered on her artistic treatment of herself as a woman, her body, and her family, all in the context of her western culture.

In studies like these, the evaluative criteria must support that aim. In critical design objects, they have to support criticality. Unlike in expansive evaluation, there is no conceptual criteria that creates the domain to be explored, and as the name probing suggests, there is no demand for being systematic. The character of probing evaluation also suggests that being systematic in the way of repercussive evaluation would defeat the purpose of being selective. By ruling out typical scientific criteria of goodness, probing evaluation also pushes criteria like rigor to the background. The main success criteria becomes novelty, surprise, and skill in executing the designs. By these standards all three dissertations described above passed the mark.

6.2 Evaluation in Methodic Tradition

When design is seen as a science, evaluation again starts with the premise that people who use an object of design will develop their own understanding of it. The way to know these understandings is to create a theoretical model of user experience, conduct a series of experiments, and then compare the findings to the model. The process encapsulates meanings into the research world, and the main success criteria for the theoretical proposition that drives the work is whether it manages to predict user experience. To make a valid contribution, researchers must use well-known methods, and make their research process transparent. Evaluation is controlled.

We prefer to talk about this method as an *accumulative* method. This method is generally characterized by the evaluation of a design experiment according to 'thickening' the knowledge contribution of one nucleus of criteria. From our analysis of dissertations accumulative appears native to the research tradition of 'lab' serving a knowledge contribution ideal of predictability. To secure as controlled an evaluation as possible all disturbing factors and contextual relationships are excluded. All insights gained through sketching, mock-ups, and prototyping during the design process are then held up strictly against the nucleus of criteria. As design iterations are performed knowledge outcomes may layer themselves as circuits around the nucleus, but the understanding of every outcome is gained only by falling back into the center, which is why we refer to it as accumulative.

This method of evaluation can be seen at work in the work of Philip Ross (2008), among others. Ross, for instance, uses a constructive design research cycle to investigate how three different perspectives on interaction behavior (Dynamic Form, Social Activity and Sensor-Motoric Activity) can inform the design of intelligent lamps. Specifically, he wanted to investigate whether it is possible to design the lamps so that they elicit certain ethical values in user's experiences. For his lamp design, Ross deliberately chose neutral colors, material and a static shape to rule out any influence from factors other than those of interest.

Through a series of experience prototyping experiments documented meticulously, Ross explores various ways of correlating dynamic form, social activity and sensory-motor levels. While this result detailed design guidelines for how the three perspectives may inform design of intelligent lamps, his evaluation showed that the lamp designs do not elicit the pursued values in user experience. Hence, Ross ended up falsifying one of his hypotheses.

The accumulative methods is consistent with the idea that constructive design research is a science. The success criteria are also familiar from the sciences. There must be progress from one study to the next (or at least a study needs to rule out a rival hypothesis), the path from one study to the next must be well argued, and the final contribution is measured against the starting point. By this standard, both works do well. Finally, the methods employed in the studies must be done using the best available methods that are understandable to others in the field. If these conditions are met, the study passes assessment with high marks.

6.3 Evaluation in Programmatic Tradition

As John Zimmerman and his colleagues (2008) have implied, evaluation can be done against the discourse behind research projects. The measure of success is essentially whether a new study contributes something new to previous research. Such revision and addition is familiar from the sciences, but also from the social sciences and the humanities. A good deal of design research picks its models from these fields, where a critical analysis without new data can be a contribution if it exposes, say, conceptual and theoretical problems in previous studies (see Koskinen 2015). The heart of the method is comparison to discourse.

Comparative evaluation happens when evaluation is used primarily to explicate and judge how design work exists in a system of family resemblances and differences in relation to the author's portfolio. Similarities and differences are relations that can be either intrinsic (much like in a designer's portfolio), or they can be external (as when design work is compared with the works of other designers). As a requirement for this method of evaluation, a core criterion must be established. It can be a key concept, term, measure, theme, of something else, but it has to be there. It specifies the nature of comparison —it is one thing to add a concept to discourse (see Battarbee 2004) and another thing to add a theme (see Busch 2008), for instance. By specifying the nature of the intended contribution, researchers can target their work to guarantee they can achieve their goals.

This method is widely used and well documented in Niedderer's *Designing the Performative Object* (2004). In her case performative objects (Art Objects, Ritual Objects, Craft Objects) act as case analysis that through comparison ground the nucleus in the model — documenting that there is such a thing as a performative object. And the design experiments (the Libation Cup project, and the Social Cups projects) are done and evaluated to point to how one might more precisely design a 'performative object.' She defines the concept of 'performative objects' as a core

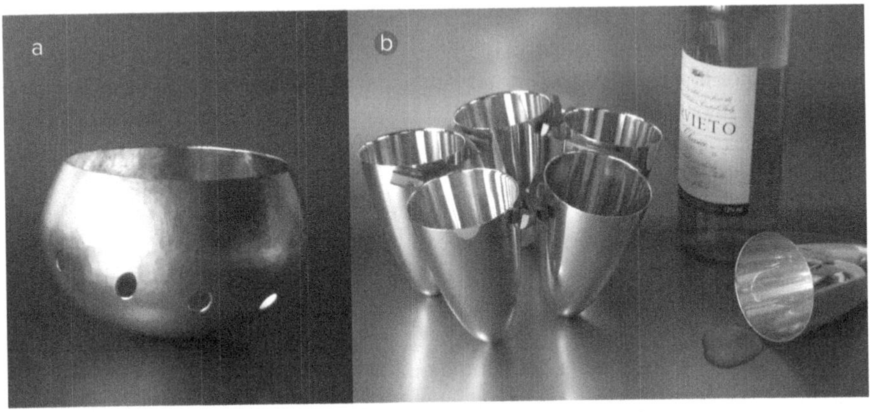

Fig. 6.1 (**a**) Libation Cup, (**b**) Social Cups

criterion for conducting an evaluation of her design work. She draws upon several theories (e.g. semiotics, hermeneutic phenomenology, behaviorist theory) in order to derive two defining sub-criteria of performative objects. Thus, performative objects are objects that are designed to (1) prompt an experience of mindfulness (what she refers to as 'result') through (2) a disruption of function (referred to as 'means'). Through a careful analysis, she identifies four object categories against which her performative objects are compared: art objects, ritual objects, design objects and craft objects. Art objects, ritual objects and craft object are able to evoke mindfulness of a similar kind as performative objects, but they do so through different means than disruption of function. Only design objects qualify in this respect.

However, her evaluation is not restricted only to the external comparisons between performative objects, and art, ritual and craft objects. Taking the center stage is her evaluation of how her framework allows her to explore some design possibilities for designing drinking vessels as performative objects. In the Libation Cup project (Fig. 6.1), she deliberately designs a cup with holes in it to evoke mindfulness-of-self in relation to the object and symbolic connotations. In the Social Cups project, cups without foots are designed to encourage users to be mindful of their 'interpersonal interaction' during the drinking act as the cups can only be placed to stand on a table when combined in sets of three. To sum up, she uses evaluation of external relations to name and classify performative objects as a new object category, while her evaluation of intrinsic relations is central for making a nuanced distinction between two forms of mindfulness.

6.4 Evaluation in Dialectic Tradition

If design is seen as a dialectic endeavor, evaluation needs to take into account how the target community takes up the design proposal. Evaluation is not a methodic exercise, as in accumulative evaluation; nor is it like in probing, which builds on the idea that the designer is the ultimate cause behind a piece, and hence evaluation happens in relation to the designer's intention; nor is it anchored to an annotated portfolio. Rather, the community is the anchor. Its response to the design becomes the main success criteria. This response can be captured methodically in many ways — ranging from short videos to field studies and perhaps even surveys — but the baseline remains the community and its response.[1]

In the corpus, there are two main ways in which evaluation can be anchored to the community. First, evaluation can proceed in an expansive fashion. It resembles a voyage of discovery where new places and insights are described along the way much like annotating bits of observation and information onto the traveler's map.

Such is the case, for instance, in work of Christian Dindler (2010), who introduced 'engagement' from the theories of Berleant, Borgmann (1995) and pragmatism as a valuable new territory for designing 'participatory engaging' exhibitions in museums. In his work, the ideal of active user involvement created an outer border for the expansive model. He excluded any experiment that would have resulted in pacifying exhibition audiences. He included all cases in which active user engagement was fostered by participation and interactive technologies. To shape what is considered the inside and the outside or the scope of the project Dindler took his point of departure in analyzing the fairly traditional museum space of the Viking Ship Museum in Roskilde, where active involvement of the audience lies at an absolute minimum. The shop served as a low-level indicator for his evaluation of two interactive prototypes, which he developed and designed for the Kattegat Marine Centre and Moesgaard Museum respectively (Fig. 6.2).

In Kattegat Marine Centre the installation allows for 'engaging visitors in playful activity' in creating imaginative fish and sea creatures, whereas in Moesgaard Museum, he created visitors a possibility of creating their own interactive runic stones. The purpose of these stones was to prompt them to reflect on seminal events in their lives and tell these to other visitors. In this way, Dindler was capable of

Fig. 6.2 (**a**) Constructing a fish table – Kattegat Centre, (**b**) setting a runic stone

[1] The best analogy is member validation in qualitative research (see Emerson and Pollner 1989).

unfolding and expanding upon the notion of engagement. His work contributed to both frameworks, and it also created theory that helps us to understand the relationship between engagement and transcendence in museums. For him, fictional space provided options for the planning of participatory design processes to go beyond existing practice, and imagine radically new situations.

Dindler's study shows one way in which designers can create benchmarks for their design experiments, and expand their work from that base. Another dialectic way to evaluate design work is illustrated in the work of Aviaja Lynggaard (2012). She was interested in so-called 'tactics for making home' she wanted to derive from a series of ethnomethodological studies of highly mobile people. One tactic was *connecting* — staying in touch with those at home over distance. Another tactic was *territorializing* — distributing one's belongings in a space to make it homely. Rather than starting out from a fixed nucleus of criteria, Lynggaard took one tactic at a time and explored it further in concrete experiments. In a serial manner, she evaluated a particular tactic for 'homing' in each experiment by comparing it to the previous experiment. At the end, she was able to present a taxonomy consisting of seven homing tactics and provides a contribution on theory of human practices in line with ideals for frameworks and categories.[2]

6.5 Accountability as Drifting Heuristic

In this chapter we have argued that contributions in constructive design research is a matter of to whom and what the research is held accountable to, and a key measure of accountability is how and what you chose to evaluate. We have argued that constructive design research would benefit from embracing and including the rich diversity of the field's research practices. In literature, several positions are dominant representing opposite opinions concerning the criteria and purposes of such a methodology. One position proposes a cross-disciplinary perspective where research is seen in relationship to models of analysis and evaluation borrowed from existing research cultures (Koskinen 2015). Another situates evaluation to the conceptual domain in design research (Zimmerman et al. 2010). The third claims that design constructive research has a unique epistemology and pushes it to develop its own standards instead of importing them from other disciplines (Gaver 2012).

We take a different route. We have argued that the best way to understand evaluation is by explicating and classifying existing research practices in the field with the perspective of how it holds itself accountable — the epistemic tradition in which you work. This is the backdrop on which the K-R model forms the core of the book. It elaborates constructive design research processes by placing them into the middle of two audiences, with different criteria — theory and impact — pointing to a

[2] The method has a precedent in analytic induction in the social sciences. See Koskinen 2003 for a design-based description.

valuable contribution. This has profound implications to research. When design is understood as a methodic (or as a putative science), researchers must hold themselves accountable to epistemic standards that drive science. When an experiment has been designed to evaluate the theory that underlies design, evaluation has a relatively straightforward function. Its role is to decide whether the theory behind design work holds water. This leads to a model in which experiments accumulatively build knowledge. The method of evaluation must be transparent and logical; if it is not, it is a source of error. Frens's (2006) use of analysis of variance is exemplary in methodic design research.

When designers see themselves as practitioners or as artists, the direction of accountability changes. Drifting can happen even in the evaluative phase, or even after it. Maarit Mäkelä (2003) treated debate and thoughts raised by her three exhibitions as data. Experiments were vehicles of imagination, not tests. The main success criteria is novelty and whether the designer's storyline leads to imaginations and new insights; logic does not matter in this process. Ultimately, this method of evaluation takes design back to the idea that the measure of quality of a design object is the intention of the designer. The story changes a little when design is seen as craft. Analogous to the artistic view, one ultimate criteria for evaluating goodness may be the designer, but there is usually more: skill in execution, the designer's portfolio, and the design tradition behind the work. These become the baseline against which progress can be measured. So, for instance, after Kristina Niedderer constructed her performative objects (2004), her work has faced three kinds of evaluations: her vision, the quality of her work, as well as its relationship to jewelry — all this in addition to research discourse. The work is accountable not only to other researchers, but also to the design community (Koskinen and Krogh 2015).

When we move between the two accountabilities captured in the K-R model, we can start to appreciate how constructive design research has to juggle between two very diverse sets of criteria. One outcome of this chapter, however, is that researchers have developed ways to manage these criteria. By knowing their preferred epistemic tradition, they can evaluate their work in ways that anticipate the expectations of their audience.

- In the experience-based tradition, their work must comply with the standards of their art and craft
- To make a methodic contribution, the rules must adhere to the standards of theory in the field, have coherent research designs, and use statistical or qualitative methods that are up to the standards of scientists and social scientists
- In the programmatic tradition, they must propose a framework that brings something new to debate; and
- a good dialectic oriented designer must show more than lip service to users and communities she works with. Her work must be informed and guided by the community, and there must be evidence of how this impact has happened.

We can give a few other pointers as well. While it is possible to interpret our argument through the dichotomy of two other cultures of science and art, which was suggested by C. P. Snow (1959), we believe that the picture is by now complicated.

It is better to see constructive design research as a response to the needs of the design world, which has its own 'third culture,' as several authors has suggested (see Frayling 1993; Archer 1979, 1995; Cross 1999; and also Koskinen et al. 2011). We deem it relevant and critically important for the field of constructive design research to participate in the language games of science by using vocabulary and terminology that is familiar to surrounding research cultures. Our model, however, suggests that the picture is more complicated and pay its dues to design practice as well.

This chapter has related evaluation in constructive design research to the methods of drifting in and between design experiments pointed out in Chap. 5 and the K-R model presented in Chap. 4. The chapter completes the heart of the book unpacking the practices of constructive design research and how these practices and ideals on contributions relate to the four epistemic traditions pointed out in Chap. 3. The following chapter points to how analysis and synthesis in constructive design research are concurrent, mutually fertilizing activities and that going from analysis to synthesis or vice versa is not a linear move.

Chapter 7
Crossing the Hermeneutic Gap

Design is often described as a synthesizing and solution-oriented discipline that deals with 'wicked problems (Horst Rittel and Webber 1973). However, experiences from practice in design research tells and shows that analysis and synthesis are not two distinct separable modes of working. In this chapter we focus on one of the main activities of design, termed variously as 'the creative leap' (Verganti 2009), 'jumping the hermeneutic gap' (Hallnäs and Redström 2006), 'frame creation' (Dorst 2015), 'building creative bridges', 'design is intentional change in an unpredictable world' (Nelson and Stolterman 2003). They are all different notions and description of the same phenomena: how solutions in design practice escapes deductive reasoning and how design proposals will need to jump, leap, build a bridge and face insecurities that stem from making suggestions for the future we cannot fully predict. We propose *drifting by intention*. It is not black magic (Tracee Wolf et al. 2006), but basic human skill (Kees Dorst 2015; Nigel Cross 1999) that is exercised and trained in educational systems and brought to high levels of sophistication in research and professional practice.

In classical natural science, medicine and other disciplines ideally based on rigor correspondence between initial research interest, experiment and means of evaluation is key to a well-executed research project — this is accountability based on well-defined measures, typically promoted by a discipline. In constructive design research, this correspondence is less strict and predictive and vary across the knowledge traditions outlined in this book. This may be a result of designers being what Cross 1999 termed 'ill-behaved problem solvers.'

In our view, words like 'leaping' and 'framing' serve as useful starting points for understanding what happens when design researchers move from data to concept, but they are not the only ways we can find if we look at research practices. These terms too easily exhibit an overly dramatic picture of design, and bias its self-image to such a degree that it endangers cooperation with other (research) disciplines (Andrea Branzi 2010; Krogh et al. 2015). What we offer here is a systematic framework and account comprised of what Schön inspiringly called 'reflection-in-action' and 'reflection-on-action', the stuff designers do — to some it appears magically,

P. G. Krogh, I. Koskinen, *Drifting by Intention*, Design Research Foundations,
https://doi.org/10.1007/978-3-030-37896-7_7

others heroically to us very mundanely. This chapter clarifies the analysis of three previous chapters by analyzing how constructive design researchers handle the move from research to concept. As in previous chapters, our hypothesis is that this movement depends in part on the researcher's epistemic tradition.

7.1 Crossing the Hermeneutic Gap

A better conceptual alternative to these heroic notions comes from a Lars Hallnäs and Johan Redström. In pursuit of the change that design is supposed to bring about, research has proven to be useful to design. It helps to understand issues and problems, and it helps to analyze their structure. Research does not, however, tell about where to find solutions to problems. This remains a design task. In their book about interaction design, Lars Hallnäs and Johan Redström describe the demands of these two activities in terms of a 'hermeneutic gap':

> The hermeneutical gap is the gap between what is actually given and what we actually design; the gap that exists between our present understanding of a given design variable and the interpretation given by the design itself, a gap between analysis and synthesis, a gap between the actual users of given things and the 'user' we define within the design process, a gap between the 'user' we define and an actual user in being. It is a gap that opens up as we draw the different design circles. It is a hermeneutical gap since it is bridged through interpretations in a process of designing; the very meaning of design is that this gap is bridged through an interpretative act in terms of a definition. This is a logical gap, but there is also a historical dimension here; a gap between what in fact is given here and now and the change of meaning that the design will bring about. (Hallnäs and Redström 2006: p. 71)

Several famous designers have construed unique ways to bridge the hermeneutic gap. Some work on wooden blocks to create interesting shapes for architecture. Some build on metaphors like the weather forecast (Barbara Radice 1993: p. 107). Some create collages (Alessandro Mendini et al. 2012: 65). Some, like the American graphic designer Saul Bass follow a research process (Bass and Kirkham 2011: 284–285). Or, as the graphic designer Michael Bierut writes, 'every really famous designer I know has a visual strategy he or she can fall back on when all else fails. One makes lovely Matisse-like torn paper collages, another makes a complicated three-dimensional model and takes a picture of it, and still another puts big black horizontal stripes on everything.' (Bierut 2007: 27).

These are examples from design practice, however. Although there is little writing about this step in constructive design research, we examine this step next in the light of the four epistemic traditions outlined in this book. The claim we make in this chapter is that these traditions shape the ways in which design researchers can move from research to concept. If we are correct, in some traditions, the results of data analysis are to be tightly coupled with design through a process that preserves the conclusions of the research phase as tightly as possible. In others, however, the path that crosses the gap can remain opaque and research gets the status of an inspiration rather than a directive.

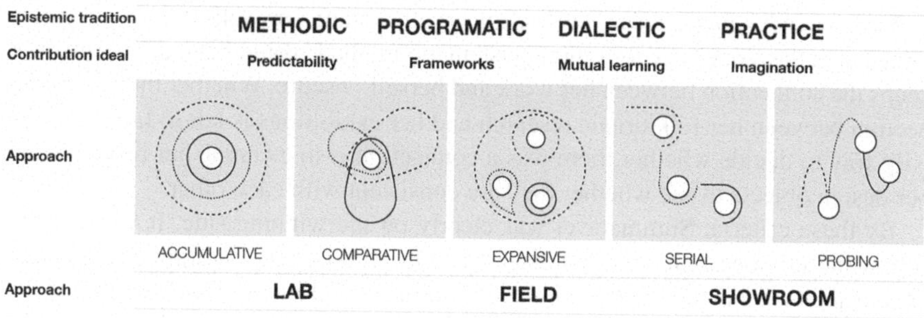

Fig. 7.1 Scheme to track drifts from design to concept

In this chapter we present a scheme that help articulate how constructive design researchers move — drift — from research to concept. As pointed to in Chap. 1 the taxonomy of Lab, Field, Showroom — places of intervention — (Koskinen et al. 2011) entice misreading leading to the conception that each is discrete and research cross borders is on the brink of derailing. As noted by the authors in the preface of the 2011 book and what is also documented in this book, this is far from the case.

The following model (Fig. 7.1) depict how the research approaches, methodological choices (ways of drifting), and the four traditions can be mapped into each other. When reading the scheme below, please be aware that the number of terms vary from line to line, and that it should not be read as ill-shaped vertical colums — there are tendencies at either end on the spectrum, yes, but no laws! Figure 7.1 is a scheme and each dissertation included in the corpus of this book provides descriptions of connections cris-crosing the map. Read horizontally the first and includes the epistemic traditions as identified in Chap. 3.

The second line describes in one word the characterising ideal of contribution associated each tradition. Line three 'methodology' includes ways of drifting. As noted above there are not direct relation vertically laws but tendencies and of course overlap as will be apparent in the following. The fourth and final horizontal line indicates the notions of Lab, Field showroom introduced by Koskinen et al. (2011). In the following analysis of the work of PhD students and their dissertations we will point to how they all to a varying degree traverse parts of spectrum indicated by the figure.

7.1.1 Experiential: The Hermeneutic Gap as Creative Space

When the concept of knowledge comes from designerly experience and assessement, researchers approach the hermeneutic gap with traditional design world techniques. There is a wide variation here. At one extreme is silence, where the designer refuses to talk about the creative process. This is the case with the Estonian professor and goldsmith Kart Summatavet also mentioned in Chap. 6. She simply refused to talk about it, even in front of the jury.

Summatavet's method can teach us many other things as well. It does not create knowledge the same way as a more systematic method would have done; rather, it keeps the connection between her work and herself obscure. Whether there is a connection between her folkloristic research and her exhibitions is left to faith, and the only way to decide whether there was a connection is to examine her research and her design objects to see whether they are consistent with each other.

By these criteria, Summatavet was clearly on the winning side. It was easy to follow her ethnographic inspired research of the folkloristic tradition and its main conclusions were clearly at the bottom of her three exhibitions. This brought her the design foundation that, for girls, Estonian traditional goldsmiths used copper, and a set of symbols that were different from the fertile age, where the preferred material was silver, and the post-menopausal years, where the material of choice was gold, and the symbols of womanhood vastly different from the girlhood. Her research specified the broad outline, but the basic storyline behind her exhibitions came from her research. The details in her enamel work were left for the artistic judgment, and even though she refused to reveal her process, her craft was immaculate and the main dimensions of her symbolic universe in the exhibitions came from her ethnographic inspired research. Like in weather forecasts, her research identified the main variables that shaped the behavior of her process, but just as meteorologists leave out microclimates from the forecasts, she left out the details that shaped her design objects (Fig. 7.2).

Summatavet's way of working resembles Jayne Wallace's method (see Chap. 3) in so far as the actual design work was of concern only to the design researcher and the motivation was to reposition its relevance as an expressional practice. Wallace based her design work on the interpretation of probes, and upon completion brought her work to the recipients of the work for comments exploring experiential overlaps between designer and recipient. Summatavet studied master goldsmiths to learn from them, with the constructive design research objective: 'to understand and visually organize the knowledge, experience and feelings collected under the influence of the fieldwork and then to 'think' visual thoughts dealing with the traditional reality.'

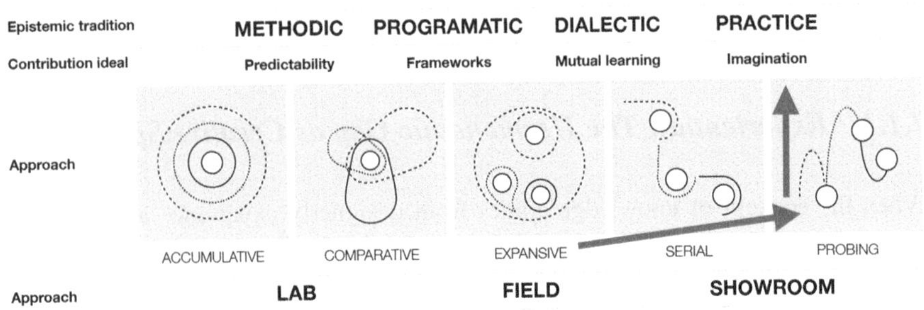

Fig. 7.2 Summatavet's trajectory in graphics

There are of course other design-world methods that can close the hermeneutic gap. Over the last few years, many of these have been fictional. For example, Bill Gaver once described his path as 'kind of gossip' followed by a series of tactics that are meant to inspire designs. They aim at identifying odd details that are exaggerated and twisted. These twisted details are turned into stories, and designs are built as props in these stories before the props are being put into a location where they meet other designs (Gaver 2002: p. slides 78–79). In *Speculative Everything*, a book by Anthony Dunne and Fioba Raby (2013) likened design to literature and described approach to design in terms of thought experiments structured as 'what if…?' statements that lead to alternative scenarios, counterfactuals that question the past, and reductio ad absurdum,[1] familiar from logic.

7.1.2 *Methodic: Making the Gap Transparent*

When design is seen as a science, the way to work though the hermeneutic gap is to use methods experts know well enough to judge whether they could be accepted in the sciences and the social sciences. A good illustration is Stephan Wensveen's (2005) carefully documented path from cultural probes to a new type of alarm clock through a design process that mostly consisted of sketching and prototyping. Probe returns gave him dream-like world of associations and metaphors that he used to design an alarm clock that would take into account the user's emotions. He moved from probes to sketching, prototyping, and finally testing his clock in laboratory-like setting with sophisticated trigonometric and statistical models. His work started from showroom-like art-inspired methodologies, and ended up in the lab-like world of logical constructs. Wensveen traversed the full scheme from right to left in Fig. 7.3.

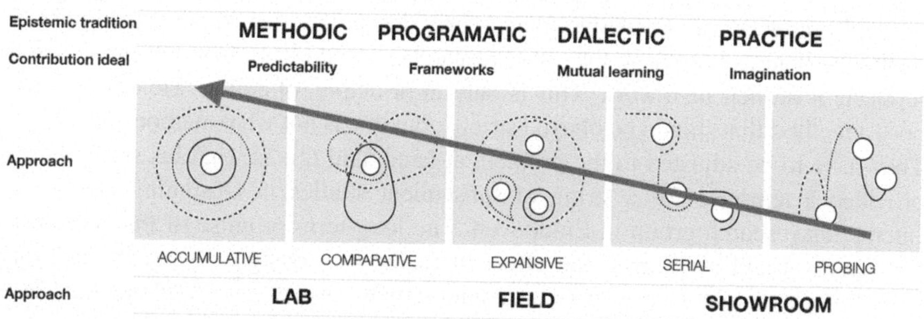

Fig. 7.3 Wensveen's trajectory

[1] In logic, one way to falsify a statement is to derive an outcome from it and then, using the same rules, derive another outcome that contradicts it.

It is good to straighten one common misunderstanding before it happens. Wensveen's work was a creative process. He got his starting point from his research, and his design process was transparent — in Bill Gaver's (2001) words, 'epistemologically accountable' — but it was creative the same way any product design process would have been. He didn't know much about the design outcome in the beginning of the process; it arose from sketches and critiques in the studio. Characteristic to his approach, however, is that any industrial and product designer can follow it, and see whether its steps were well laid down. The process preserves knowledge gained in the research phase of the study, and takes it forward. In this sense, it is like analysis of variance in psychology: the method is known and it eliminates errors coming from methods (this is known as method error or method variance in research). It eliminates one source of possible problems from research.

Behind Wensveen's way of dealing with the hermeneutic gap lies a worldview his research community shared when he was doing his work. It proposed a relatively tight connection between research analysis and design analysis and was driven by the bold pursuit, also quoted earlier to '...construct mathematical models of actual interaction.' (Overbeeke et al. 2006: p. 64–65]). Based on the assumption that proposing solutions is a matter of following an analytical research process that is working from data to design and is ultimately theory-driven.

Other researchers in the Dutch research community have taken a more moderate approach, but still have accepted the premise that design can be seen as science, or at least proto-science. For Pieter Jan Stappers (2007), for example, as elaborated on in Chap. 4, design is a process that creates a physical hypothesis. Caroline Hummels has been more skeptical about the need to build contextuality into the models that guide for design. She wrote:

> As designing has become essentially contextual, scientific work and results should also be contextual, i.e., lead to conditional laws instead of general laws. In my opinion, the classical highly controlled experiment, which tries to uncover 'the truth' through falsification, usually kills both context dependency (looking for general rules) and subtlety (control and repeatability). Therefore, we need to expand our research methods: incorporate the context in an experiment and allow for subtlety. (Hummels 2000: 1.27)

Given this concept of knowledge, Wensveen's design practice was designed to keep method variance in check so that other researchers could learn from his process and replicate it in their own work. This is only in principle, of course. He did not propose a method that should be obeyed strictly, but this is not what methods are about. They have to be adjusted to the problem at hand, and this is what his study did: it provided a template for a string of subsequent studies in Designing quality in Interaction research group in Eindhoven. The long-term promise of this approach was that it could minimized the effect of method on designs — which, in turn, would minimize one source of error in a journey to Overbeeke's mathematical science of interaction design.

7.1.3 Programmatic Tradition and the Hermeneutic Gap

Another concept of knowledge in constructive design research that has its origins in the research world puts premium on programs rather than methodic precision. The idea is that knowledge is created in discourse, while precision in methods gets a secondary position. Should a researcher follow this line, He/she does a literature review, thinks what would best contribute to that literature, creates a concept, and does his/her study. The study initiates debate and new lines of research that correct possible errors that set boundaries to the design concept, and that way guarantees it makes a contribution. This is clearly the case of many of those frameworks that had been created in interaction design. They are not testable as such, and they as a rule are formulated intuitively. Rather, they remained an unspoken-of philosophical background.

In *Shape-changing Interfaces* (2015), Majken Kirkegaard Rasmussen studied shape-changing interfaces. Her main focus was on furniture augmented with sensors and servomotors. In response to human input, the furniture she built changed shape. She installed her designs — in particular a shape-changing bench — into a settings like an airport, shopping mall and concert hall to see how people would respond to the experience of interacting with furniture that seemed to have its own will. Most of her studies were Wizard-of-Oz studies in that the behaviour of her objects were controlled by researchers rather than a program or artificial intelligence. Yet, they provided her with a method for going beyond her literature program, which in effect became her zero measurement against which she could measure the success of her design experiments.

Her research started with a careful literature review that tied her work to human-computer interaction and specifically a small strand in it that focused on shape-changing interfaces. Her literature review distinguished her topic, shape change, from robots and dissected the materials and technologies that had been pursued in earlier work. She also extrapolated user experience hypotheses from literature. These were published in a paper in (Rasmussen et al. 2012) that became the starting point for her research. The main contribution of the study was the framework she built to guide the design of shape-changing interfaces rather than the devices as such.

The devices in their part were left rough to signal they were intermediary exploratory prototypes rather than products. In contrast to Wensveen's alarm clocks, she did not try to turn them into stimuli in research experiments. In contrast to Summatavet's jewelry, her design objects were not assessed by professional furniture or accessory designers either. Instead, they gained their significance from research discourse, which became the benchmark of success. Did she advance knowledge of shape-change in interfaces? No doubt; her study became the benchmark in the area for a short while at least. If we describe her practice graphically in relation to the previously introduced scheme her overarching trajectory went from comparative literature studies and small-scale experiments to probing the context of interest building prototypes in serial and expansive manners and finally running comparative studies of the objects used in contexts (Fig. 7.4).

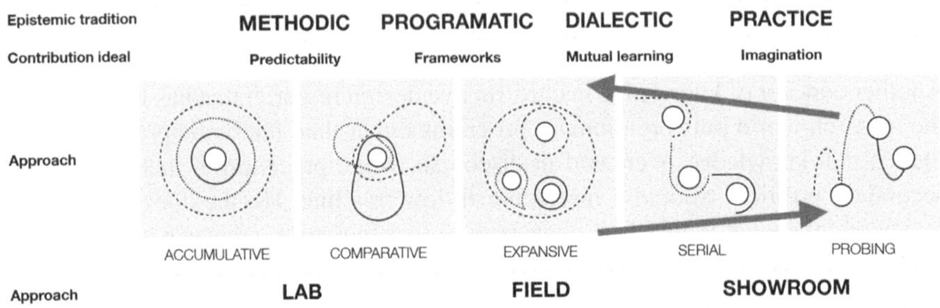

Fig. 7.4 Rasmussen's trajectory

In terms of her overall methodology, the way in which she approached Hallnäs and Redström's hermeneutic gap resembled Wensveen's, but also significantly differed from his study. The methods she used were standard design methods, and passed the assessment without much notice. The concept, unlike Wensveen's, came from very specific set of literature, and ended up with it as well. In contrast to him, the gap between research and concept in her study could be much wider because the validity of her study did not rest on method and Overbeeke's mathematical concept of knowledge. Rather, the criteria of success was whether the research community that can examine possible biases in her work to see whether it is credible. The method contributed to discourse rather than furniture design. A study like hers may lack a measurable outcome, but can make a significant contribution.

7.1.3.1 Dialectic Approach to the Hermeneutic Gap

The dialectic tradition puts premium on participation and sees it as one of the most important aspects of design. In this view, the hermeneutic gap is traversed with people designers are working with. They are brought into the design process even in that tender stage in which concepts are being created. They have at least a say, if not veto, in at least those parts of the design process that do not presume design-specific technical skills. This view is in some ways a more radical version of the programmatic view in the sense that also both put premium on the community. The main difference is the nature of the discourse. In the programmatic tradition, discourse is limited to those who know design literature. In the dialectic tradition, researchers are only one part of the conversation that defines how a design concept arises from analysis.

For example, in Andrea and Marcelo Judice's theses (A. Jucide 2014; M. Judice 2014), research started with cultural probes, which were embedded in an ethnographic study in the village of Vila Rosario in the state of Rio de Janeiro. They produced their designs in the studio, but their process built on a constant feedback from the villagers. Marcelo and Andrea organized a series of workshops in which they designed their 'products' with the health agents in Vila Rosario. In these conversations, they defined which products to create, and generated designs with them.

They also designed visual elements used in the products. The products became joint creations rather than showpieces of the designers' skill. The essential determinant of how research turned into products was co-design, which took place in conversations in the workshops. In a graphical representation of the work there is a lot of criss-crossing between serial and expansive models of drift. And as the context and the change of practices are the grounds of evaluation and justification the project stays very much in the middle of the graphical model (Fig. 7.5).

In a co-design process like this, the local community decides what is valid. This leads to designs that work in the context in which they have been designed rather than showcases that can be exhibited to design juries. The approach is different from Summatavet, whose decisions about design remained private but whose designs were juried by experts. The approach is also different from the idea that a good design builds on an explicit process, like in Wensveen (2005).

If the design has its origins in a co-design process, issues like replicability become secondary concerns and matters of curating conversation around emerging designs rather than matters of method or protocol. The outcome is a design that is robust and fitted to a particular community. It may well be that a participatory approach finds its natural home in community design (Meroni 2017) and in social design in which the unique qualities of the community tend to be key drivers of design, and where theoretical approaches — like ecological psychology behind Wensveen's (2005) work — are too universalistic.

This may sound dubious because it takes the control of the design research process away from designers and researchers. It is well in line with a long trend in philosophy of science, which had turned into real-life studies of how science is made, however. Study after study, philosophers and sociologists of science have shown how much negation goes into facts (for instance, see Bruno Latour and Steve Woolgar 2013; Lynch 1993). The argument that dislocates decisions from professional frameworks and categories to the language of the target audience of design has been presented in design through Wittgensteinian philosophy (Ehn 1988), Latour (Binder et al. 2011) and symbolic interactionism (Battarbee, Katja and Ilpo Koskinen 2005), among many others. What is distinctive here is the idea that validity seen as the property of a community — late Wittgenstein's 'forms of life' in his *Philosophical Investigations* — in which concepts become intelligible. Just like in

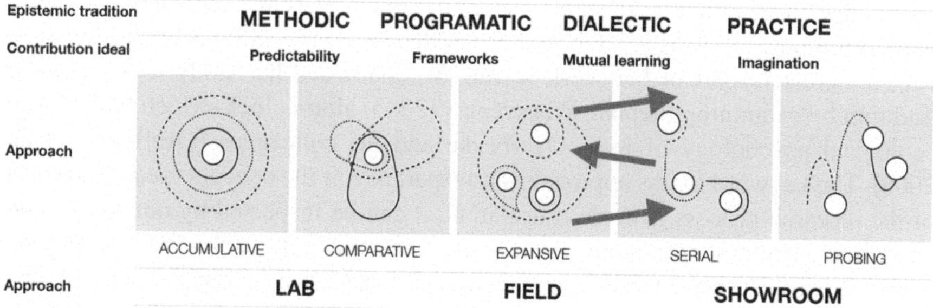

Fig. 7.5 The trajectory of M. Judice

the sciences, this implies that many concepts usually seen as precise are in fact incommensurate. As the example of the notion of 'mass' carrying very different meanings in diverse disciplines. The logic, however, extends to non-scientific communities as well. The criteria for knowledge vary from one field of research to another, and design is not an exception if we follow this line of thought.

In the beginning of this section, we mentioned that this view is in some ways a radical version of the programmatic tradition. Just like it, it puts the premium on discourse. Yet, it differs radically from it too. The co-design approach wipes out the boundary between experts and lay people in discourse. Importantly, it is well in line with empathic and participatory ideas that have been suspicious of elevating expertise to a leading position for a long time. The dialectic approach emphasizes this distinction.

7.2 From Research to Concept

Constructive design research can drift from research to concept and close the hermeneutic gap at the heart of any design process (Hallnäs and Redström 2006). We have pointed to how constructive design researchers go full circle from hypothesis construction, experimenting, and evaluating. There is a massive literature on creativity and concept design, but this phase has not been a popular topic in research. Descriptions of how famous designers and well-known companies traverse the gap exist, but they are anecdotal and often in the form of second-hand descriptions and are short in detail. The literature also is dominated with the ethos of Silicon Valley which equates the value of design with radical and disruptive innovations, which is by no means the only possible way to travel from research to concept. These descriptions tend to paint an overly dramatic picture of design as a heroic discipline leading to dramatically new ways of seeing the world. Our proposal is to see the transition from research to concept as a drift rather than a jump or process that can be described formally and optimized.

Constructive design researchers have approached this crucial creative step of their craft. The hypothesis of the chapter has built on those four epistemic traditions we identified in Chap. 3. We started our exploration from a tradition that located knowledge in experiential design practice by studying Kart Summatavet's thesis, which built on a folkloristic study and proceeded as a series of three exhibitions, but kept the creative part of her work secret. We contrasted her study to a methodic tradition by examining Stephan Wensveen's (2005) alarm clock, which built on the ecological psychology of Kees Overbeeke and his colleagues (Overbeeke et al. 2006). The keyword to this approach is transparency: if the concept creation section of the research process is made transparent, it can be inspected by outsiders, who can judge whether it is reliable and at least in principle replicable. We then studied a third approach, which built on the idea that design research happens in discourse — is programmatic. This approach is exemplified by Rasmussen's (2015) work on shape-changing interfaces. Finally, we examined the dialectic tradition, in which drifting is justified by people, and illustrated this with the theses of Andrea and Marcelo Judice's theses (A. Judice 2014; M. Judice 2014).

One implication of our exploration is that there is a plurality of ways to approach the hermeneutic gap. More importantly, if we are correct, there is an underling order in this plurality. The way to traverse the gap depends on the underlying concept of knowledge, which explains some of the possible controversies in the field. Some approaches are clearly incompatible with each other. It is hard to imagine how Summatavet's approach could be accepted in a research culture that sees design as a methodic. It is equally easy to see how Wensveen's and Rasmussen's approaches would be found uninteresting in the *experiential* tradition, which sees top-notch design work as the main hallmark of quality.

As a whole, this chapter has explored a territory that has received relatively little attention in constructive design research. Still, the step this chapter has focused on is crucially important for it. To make constructive work interesting, design work has to be made interesting with an intellectual scaffolding. We are fully aware of the fact that our exploration in this chapter builds on our interpretation of epistemic traditions in constructive design research, and we do not claim that our interpretation is the only possible way to approach the hermeneutic gap. We have, however, explored the gap and we hope our exploration will breed discussion in the future. Our contribution, we believe, lies in the way in which we have posed the question of how design researchers can translate their research into concepts. We claim that this path crucially depends on the underlying concept of knowledge.

Chapter 8
New Experimentalism

This book has argued that the heart of constructive design research is usually a design experiment; it is a key vehicle of knowledge creation. In this chapter we study design experiments through a fairly recent philosophical debate that has been targeted against the theory-centric view of research. It is called 'new experimentalism,' which argues that experiments often create knowledge regardless of theory. Following new experimentalism, the hypothesis we explore in this chapter is that design experiments generate knowledge, but also that there are several kinds of experimental protocols in constructive design research.

While philosophers of science have usually described progress in research in terms of dialectic between theory and empirical research, they have also usually given priority to theory. Ultimately, as the philosopher of science Imre Lakatos (1974) has argued, there are no such things as decisive experiments: theorists always win because theory is more flexible than experimentation. The logic is easy to understand: setting up an experiment is costly and may require massive equipment. Theory is cheap in comparison. Theory is easy to adjust compared to making another experiment to respond to theoretical criticism.

This view is not without its critics. One well-known problem with these philosophical ideas is that they usually build on physics, which has served as the main object in the philosophy of science all the way until the end of the 1980s. Its most celebrated books (Thomas Kuhn 1962; Imre Lakatos 1970; Bunge 1977) mostly build on two cases: Newton's theory of gravity, and Einstein's theory of relativity. These are among the crowning achievements of science, later developments in philosophy have questioned whether they can be used as models in disciplines like biology and economics. As we already have hinted at in the beginning of this book and in the previous chapter, disciplines like ethology, geology, or anthropology can be fitted to this model only with grave difficulties. This has not gone unnoticed in recent philosophy of science, in which the cutting edge has left physics and moved to disciplines like philosophy and economics after the eighties. The resulting picture of research is much more complicated than during Cold War, when physics received lavish funding because of its military importance.

P. G. Krogh, I. Koskinen, *Drifting by Intention*, Design Research Foundations,
https://doi.org/10.1007/978-3-030-37896-7_8

In arguing for 'New Experimentalism' the American philosopher Deborah Mayo (1996) has noted how the philosophy of science followed for decades so called logical empiricism that saw experiments as unproblematic vehicles of verification (and later falsification) of theory.

When Thomas Kuhn pointed out paradigms in *The Structure of Scientific Revolutions* (1962), experiments started to get a far less important role in philosophy, which instead got interested in theories and worldviews. Against this shift, new experimentalism argued that the aims of experimentation are varied, and often independent of testing or filling out theories; that knowledge created in experiments may live on without theory; and that observations may be valid without theoretical backing. Experiments are done for lots of reasons, not only for verifying small, topical theories, and certainly not only for global changes of world view using the terminology of Ian Hacking, an eminent philosopher of science. Experiments are stubborn and tend to stay uncontested even when theories change.

We wrote this chapter in part because design research, especially in Scandinavia, has had a love affair with the concept of experimentation, which we find difficult to accept without a long footnote, which is this chapter. What we try to achieve is showing that we may need to get rid of scientific notions of experimentation to truly get at what experiments may be about in constructive design research. The problem is not trivial. Experiment is a scientific term, but it is used in design practice (and in art) in a lay sense. Yet, as soon as it is brought to constructive design research, it tends to mislead designers into long and winding side paths of methodology. We believe there would be terms more native to design telling of the shape of an experiment, like exhibition, field testing, and critique, so we see the following pages as an intermediary step towards a broader understanding of design as *Wissenschaft.*[1] Design is certainly a means for acquiring knowledge. The word experimentation, however, is by now a part of design discourse, and needs clarification.

8.1 How Knowledge Depends on Experience

Logical empiricism lives in some design world assumptions. At least since the design methods movement (J.C. Jones), design researchers have been formulating better design processes and mapped methods to these processes. The dream is to build a process that would take the individual element out of design. With a perfect process, the personality of a designer would not matter: take any designer, and you get the same result. This culture has always — and probably always will — targeted an opposite culture in which designers are seen to be personalities. As a rule, the methods movement tried to push the personality-centered view aside. It met resistance form another tradition that prevails in the art and design world in which

[1] We use the term *Wissenschaft* here to point to design being a research practice concerned about making and deriving knowledge rather than posing a body of knowledge as inscribed in the use of the term *science*. We will return to difference in the following concluding chapter.

personal character is seen as a prerequisite for design quality and identity. Constructive design research has a foot in both camps and tends to oscillate between them.

We believe new experimentalism gives us a way to rephrase the relationships between these traditions. Above all, it suggests that design experiments have a role in knowledge creation. If scientists and philosophers of science are fine with the idea that experiments have a life of their own, and can lead to new knowledge, the claim that design experiments do that should be equally well grounded. This observation may be obvious when we look at recent research, but with the possible exception of the methodological treatment of Koskinen et al. (2011), they tell us little about what kinds of processes are involved in turning design experiments into knowledge.

If we take Pieter Jan Stappers's spiral (see Chap. 4) as an example, this model points to both the general scope of a project and that knowledge enters and is disseminated from the project. Central to the graphics is a white spiral-like arrow depicting iteratively moving processes that integrate a variety of knowledge supporting the overall structure. Equally importantly, however, he draws smaller red arrows pointing into the spiral illustrating that knowledge enter from the outside and red arrows pointing out of the spiraling movement signifying that knowledge is generated and disseminated from the acts of designing. What he does not say clearly, however, is how this happens. The best guess is that — in his words — doing design generates new insights and ideas, helps to verify some and falsify others, and that way design creates knowledge. This is true for sure, but poses the question of how the process happens.

Authors like Bowers (2012) have gone back towards design with their concept of annotated portfolio. They have argued that taking design seriously as a research instrument requires that we focus our work on the design object, put theoretical ideas to the background, and generate knowledge slowly in a bottom-up fashion through annotated portfolios. Such manifestations are an attempt to provide means for documenting design concerns and decisions otherwise often hidden in the design process. Others find the answer in experiments related to frameworks and programs (Binder and Redström 2006): in their view, new knowledge is generated in discourse.

If we put the matter in these terms, we see better what new experimentalism can offer to design research. As Deborah Mayo notes, philosophers of science after Kuhn have generally 'gone bigger' in search of the source of knowledge. As she writes, to appraise a theory, progress requires considering large units like paradigms and research programs. As practically any university student learns today, observations are 'theory-laden,' which is to say that testing ideas cannot be done apart from testing larger units like theories, programs, or paradigms.

This view, as Mayo notes, is dominant even among those philosophers who try to refute Kuhn. Yet, there is another alternative that argues for taking the details of experiments seriously as a form of learning:

> The fact that theory testing depends on intermediate theories of data, instruments, and experiments, and that data are theory laden, inexact, and 'noisy,' only underscores the necessity for numerous local experiments, shrewdly interconnected. (Mayo 1996: 58).

If we take Mayo's basic message seriously, it gives us a clear hypothesis. We need to look at design research practice in detail to understand how design experiments can be instruments for knowledge production. Part of the answer lies in frameworks and programs, but another equally important part lies in those activities that Stappers tries to capture with the arrows that enter the spiral form outside theory. What they are, how they work, and what kinds of things these activities invite for.

8.2 Against Method: *Methods as Achievements*

There is one PhD thesis at the very edge of our corpus that has posed this question in the specific context of cross-cultural design, Jung-Joo Lee's *Against Method*. It pays to note that the title of her thesis was not an allusion to the philosopher Paul Feyerabend's similarly named classic in philosophy. Instead, Lee's thesis captured what she learned in her study, which was about whether design research methods are portable from one culture to another. The practice of adopting Western methods to other parts of the world is wide-spread in Asia in particular. In Asian universities and industries, it has been customary to take methods from the West and then adjust them to fit the Asian context.

Lee's approach was different. She posed the question of what it is in any local practice that makes methods work. Her analysis uncovered several interactional practices that made things like cultural probes or usability heuristics not work. Her conclusion was that turning things into methods with a major M is wrong; hence the title of her dissertation. However, Lee was mostly interested in interactional practices behind methods, joining a long line of ethnomethodologists who have studied science, and conversation analysts who have studied how surveys are made to function in conversation, and what kinds of assumptions make it possible to social researchers in the first place to impute numbers on issues like political attitudes. For our purposes, Lee's analysis is too fine-grained, but her message is well in line with what Mayo and other proponents of new experimentalism are saying.

Having said this, we need to stress that we find Lee's analysis credible. She shows in considerable detail how design methods are defined in their context of use. Her analysis is specifically about methods that most people find unproblematic, starting all the way from focus groups and usability walkthroughs. If we build upon her analysis, we can start to see where Stappers's red in and out going arrows come from. They are not formulas, and they are not well-agreed methodic protocols. Rather, they are results of local negotiations e.g. among stakeholders based on insights developed in the project and improvised solutions to problems at hand. If this is the case, it is highly likely that they can never be captured exhaustively. A good deal of learning from these practices is never written down either — after all, Lee uncovered some of them, but it took a PhD thesis and a dense sociological framework. A practical margin always remains in research, and it cannot be done away with. It can be assumed away, no doubt, but it is still there, making some things easy and others recalcitrant.

When she turned design methods from a resource into a topic, she came to show that even though they may look clear to outsiders, they in fact depend on many types of practical issues. For her, design practices and traditions proved to be stubborn: they are not something that can be done away with by improving methods and processes. How they work depend on the actual practices that keep them running. By studying them, we can learn a lot.

8.3 Design is Stubborn

We are finally in a place in which we can start to see how to understand design objects in the heart of design experiments, be these, art works, spaces, interactive systems, services, and social organizations. The cue we build on again comes from new experimentalism. Mayo (1996: 271) quotes the Stanford philosopher Peter Galison, who has noted that: 'experimental conclusions have a stubbornness not easily cancelled by theory change.' (1987: 259) (Quoted in Mayo 1996: 271). This goes against Lakatos's claim that ultimately, theorists always win, and no doubt resonates with design professionals.

We believe that this idea fares even better in constructive design research, which has never cut its ties to practice. We have seen how this field of research has generated a proliferation of frameworks. These frameworks, as we have seen, are often reflective explications and, as such, afterthoughts rather than theories. They, however, become theories if other researchers pick them up and use them in their studies as arguments. Constructive design research has generated by now a world of potential theory. The problem is: how do these frameworks relate to the things researchers generate? Are they central to research, as those writers who locate progress in the conceptual scaffolding around design work say (for example, Zimmerman et al. 2007), or are they a residue, as other views seem to imply (for example, Gaver 2012).

This question, as useful as it may be, is obviously too categorical to be of real value. Both arguments have merits, so the real question is how theory and experimentation work in design? One feature of constructive design research is that a good deal of it proceeds through previous design cases and frameworks that externalize knowledge from them. It is rarely driven by statistical analysis, and certainly not by theory only. Instead, the act of designing, making design moves is the fuel for development. Looking only at the final product one will lose much of the knowledge created in designing objects.

An old piece of social psychology provides a useful case for understanding how the loss happens. Attribution theory has been influential in social psychology since the sixties. Its basic claim builds on an observation of Harold Kelley from UCLA. Its research object is the ways in which people attribute feelings, intentions, characteristics and so on to other people, and it has led to thousands of useful studies about how these attributions are made, their correlates, as well as their consequences to issues like age, ethnicity, and gender. It says that the way we see human action depends on our perspective. When we walk on the street, talk on a phone, or make purchases on an expo, we know not only what we are doing, but also the background: calculations, hesitations, alternative routes of action, and doubts of many sorts. When we look at someone else doing the same things, we do not see the background, but the action. If we have to explain why other people believe or do something, we tend to see order and decisiveness. If we turn to ourselves, we see hesitations and choices. We attribute reasons differently depending on our perspective.

If we take this piece of theory, and apply it to design research, we see a similar difference. When we look at our own work, we know the alternatives, and tend to see the work as quite a bit more fragile than outsiders, who do not know the background. The only thing that gives them access to the background is what we tell them. Vice versa, when we look at outsiders' work, we see more intention and order than they see. In case we want to understand a piece of design, our reasoning depends on our perspective. In design critique, the methods are similar to art criticism: learned, and often based on tradition, theory or even ideology based as design history will show. In industry, the methods are stories scrambled together by industry watchers. In research, they take the form of teardowns and product deconstruction. These stories may not be true, though.

Regardless of the method, though, design remain stubborn and, ultimately, escapes definition. A good case is *UTOPIA*, the best known participatory design project, which followed Harry Braverman's sociological insight in deskilling and tried to develop ways to salvage workers' skills in graphic industry at the face of the digital revolution. When one of the main researchers of the project, the Swedish information scientist Pelle Ehn, wrote his PhD thesis in 1988, one of his questions was why the low-tech methods of *UTOPIA* worked better than computer science diagrams. His interpretation came from Ludwig Wittgenstein's notion of 'language games': with low-tech things like cardboard mock-ups, designers were able to create a joint language game with graphic workers. This eliminated a good deal of the attribution problem: when both parties develop interpretations together, they share the same attribution model,.

Two decades later, in a co-authored book *Design Things*, Ehn's explanation for the success of the methods came from Bruno Latour, a semiotician and ethnomethodologist rather than a Wittgenstein, natural language philosopher (Binder et al. 2011). The logic that underlies these explanations is a proof for new experimentalism, we believe: design creates knowledge in ways theory cannot. If theory 'underspecifies' design, as Gaver (2012) says, design knowledge must be complemented by doing design.

8.4 Discourse is Equally Stubborn

Until this point, we have followed Mayo's new experimentalism to see how far it leads us. It gives us a powerful way of seeing how design experiments can generate knowledge. It has problems, of course. Most notably, with this theory, it is difficult to avoid bringing in creeping assumptions about design as a science. As we have seen, this is a troubled notion, and fits only to some versions of constructive design research. Another problem is that this description does not respect many — perhaps even most? — research practices in constructive design research.

Having said this, the notion of 'experimentation' is useful in constructive design research in several ways as it serves to be an umbrella for a rich variety of forms and allows for a diversity of detailing of the design artefacts. It includes e.g. the notion of low and hi-fidelity prototyping (see Säde 2001) that is well established in industrial design, human-computer interaction and interaction design. Low-fidelity describes a design experiment being executed in a very simple (brute) way with limited or no functionality while a hi-fidelity prototype describes a prototype close to an envisioned market product or system with or almost all features and characteristics implemented.

A design experiment may be both. When we talk about a design artefact as an experiment, it becomes an object that is meant to study an idea, and tell whether this idea is promising or not. By implication, the term assumes a discursive background that gives meaning to the object, and tells that this background is where knowledge creation happens at least in part. This is the intuition that has been driving those design researchers who have stressed the importance of discourse, no matter whether the discourse takes the form of frameworks or programs (like Forlizzi and Zimmerman), and no matter whether these terms are defined in architectural or philosophical terms (like Binder and Redström).

Although new experimentalism, then, has opened eyes to seeing that experimentation as such can, and does, generate knowledge, it should be taken cautiously as a guideline. Those 'bigger units,' as Mayo has called them, have not disappeared anywhere. Frameworks, theories, and programs remain a fact we need to understand, just as we need to understand that experiments lead to many types of knowledge that these 'bigger units' do not really care about. Going all the way to the conclusion that experimentation is everything that is needed would be like concluding that since it is impossible to achieve a 100% aseptic environment, we as well could do — or even should do — surgery in a sewer, as Robert Solow, the Nobel prize winner in economics has quipped (see Clifford Geertz 1973: 30).

8.5 Design Practice, Programs and Discourse in Empathic Design

If this analysis is correct, constructive design research has two stubborn sides, design and discourse. Although the way in which they work probably varies from one research program to another, neither side can be forgotten in a fruitful research program. To see how they work in unison, we may turn back to empathic design, an interpretive design research program in Helsinki.

As described by Mattelmäki et al. (2013), it started as an answer to a question the digital industries were posing at the end of the nineties: how to design when the material of design are bits rather than classical physical material? Old methods from design worked to an extent in this new world, of course. Industrial designers were able to work on physical shapes and on some aspects of interfaces; textile designers on fashion aspects of things like mobile phones; and design managers on design processes needed for these vastly complicated high-tech objects. Yet, the crucial question at the bottom of the industry was beyond the methods of design: what to design when you have technology that can do almost anything?

In Helsinki, the answer was empathic methods that gave designers a way to know people, turn this knowledge into understanding, and turn this understanding into proposals designers and engineers could use in their work. The program started with questions about methods, and culminated in *Empathic Design* (Koskinen 2003), and three dissertations, Katja Battarbee's *Co-Experience* (2004), Tuuli Mattelmäki's *Design Probes* (2006), and Esko Kurvinen's *Prototyping Social Action* (2007).

When the methodic questions were by and large settled, the attention of the next generation of research went to studying how to introduce understanding gained through empathic methods into industry. The answer came from various co-design techniques and design games (Mattelmäki 2006; Vaajakallio 2012), and from studies of domestication aimed at improving designers' understanding of design in everyday life (Paavilainen 2013; Ahde-Deal 2013; Peteri 2017). When Silicon Valley popularized the idea that only radical innovations are worth doing around 1995–2000, and industry started to accept creative methods a few years later, the program's response was to bring in artistic techniques like exhibitions, and conceptual ways of thinking from art. This work culminated in Mattelmäki's *SPICE project*, and relationist aesthetics at the bottom of Yiying Wu's PhD thesis *Bicycles and Plants* in 2017. The program kept its core beliefs intact, but shifted its attention. Whenever there has been enough evidence that current ways of working are wanting in some respects, empathic designers have posed new questions and sought ways to respond to these.

It is important to note that every single study in the program was done through several drifts all along. Battarbee, for instance, started with studies of interactive toys, went on to study and build interfaces and virtual prototypes, and finally settled to the concept of co-experience, which she elaborated into a theoretical statement with Herbert Blumer's version of symbolic interactionism. Mattelmäki, in her part, worked on cultural probes as early as 1998, but then ventured through technical studies of probing into a series of field studies of their usefulness in industry. The third constitutive study of the program was Kurvinen's (2007), which drifted from studies of mobile multimedia messages into a methodological statement about how to prototype social actions. Similarly, when industry and students wanted to know how empathic design turns into physical devices, the response of the program was constructive: it built an electronics lab that generated numerous prototypes, several company links, and also a link to traditional materials like textiles and plastics studios (see Mikkonen 2016).

As this case illustratesshows, design experiments like prototypes, no doubt, are stubborn, but this is not the whole story of research. The artefacts produced in constructive design research processes can be interpreted in many ways, and it is this ambiguity that makes their status as research objects risky. If the way in which we see design is completely up for grabs, it would at worst give raise to a cacophonic research community without much chance of reasoned debate. There is a real need to create frameworks that help both designers and their audience to understand how the things came about. It is the cumulative power of all this work that makes empathic design versatile and gives it flexibility in the face of shifting interests in industry (and to a lesser degree, the research world). Perhaps the secret lies even more in discourse than in design: discourse works as memory, serves as intelligence identifying new questions and directions, and clarifies how to best respond to emerging problems.

To put it in simple terms: each design artefact in the empathic program finds its significance from the research program. It finds resources from precedents within the program and it contributes to it. Without the program, a project like Wu's *Bicycles and Plants* would be almost a random contribution to — say — service design, social design, aesthetics, design anthropology. It would have been impossible for her to say what kind of argument she is making, and what kind of new knowledge she is generating. By implication, she would have been open to almost any criticism coming from outside — and constructive design research has seen more than enough of bad, unfounded and ill-willed criticism. When working in a program, the contribution is much easier to identify: methodically, she extended the artistic explorations of the 2008–2011 era and elaborated Acharya's ethnography; topically, she extended the service design part of the empathic program that started with SPICE project into social design, and she extended Kurvinen's argument about the need for developing methods for prototyping social action; and conceptually, she elaborated Koskinen's (2016) argument about aesthetics in social design. Working in a program, in short, provides the benefit of massively reducing the interpretive space any piece of design has to face. Programs are handy devices for managing the interpretive flexibility of design.

Again, this finds a design world counterpart. Theories may be alien to it, but precedents and references are not: whenever design practitioners are faced with a new problem, they start from understanding current solutions, posing questions of why they work, and why they do not work. Their research pays attention to many things design theory does not talk about, like kiln temperatures; how to spread a nano layer of salt on the surface of ceramic during baking; how to develop knitting techniques for interactive materials; fire safety; how to find cheap solutions that still will enable a rich variety of instances of a product; how to set up exhibitions; how best to create joints for office furniture; how to make a curve aesthetically pleasing; and so on. It is a secondary, that these studies target existing products rather than proceed through theory. Conducting studies in the material world is different from conducting studies in the world of research literature, but it is a context all the same. Advances can happen through theoretical work, but also through as a series of barely noticeable changes in things. Obviously, designers do not share everything they know, but they share a lot in a collective effort to up their occupation: technical skills and other tricks of the trade.

8.6 Experimentation as a Site of Intentional Drifting

For design researchers, new experimentalism is in some ways a treasure trove of questions. When we see design practice through this philosophy, we come to see it as a source of knowledge. The beauty is that we do *not* know what kind of knowledge an experiment generates before we do it. The problem with notions of experimentation inherited from philosophy of science is that they reduce the significance of experimentation and observation to theory. If this view was accepted in design, this would massively reduce the scope of things experiments can teach: experimentation is a means to an end, which is theory. Whether these theories are devised to be small in scope — like frameworks — or deeply penetrating systematic theory is a secondary issue: with this thinking, we would be back in Kees Overbeeke's dream of design as a mathematical science.

In contrast, new experimentalism opens up another way of seeing experiments as a source of knowledge: anything in design practice can become a source of knowledge. This corresponds much better to design intuitions. If we look at the line of Finnish chairs from Yrjo Kukkapuro to Ilkka Suppanen and ditto Danish from Jørgen Gammelgaard to Kasper Saalto. Welding has developed; knowledge of machines for making parts has improved; use of materials has expanded; knowledge of joinery has improved; so has knowledge of attachments. The classical development of archetypical design objects like chairs follow a criteria of novelty: new materials adopted for furniture production; new ways of production reflected in joints and assemblies; new functions; unmet customer needs; and new business models. If none of such is embedded in the product, it is considered only a matter of fashion statement in the design community.

Progress in this field of design takes many routes and cannot be judged in terms of whether every new theory confirms of rejects earlier theories. Few chairs owe their existence to theory. In fact, if we put it this way, the whole idea is ridiculous and serves to show that there is something wrong in treating design as theory. The idea that these chairs would change our global worldview would be optimistic, to say the least.

Of course, this is also the case in any serious production. Products get better in many ways simultaneously, especially in mature industries. (Look at mobile phones or cars). This is also true of design research — when we do design in research, its aims are different, but it should still function as design functions.

New experimentalism contains a very radical, almost anarchic theory of drifting. It opens doors to several kinds of drifts simultaneously. If we take our cue from new experimentalism and leading design practice, we must learn to pay attention to detail and make better experiments through a series of small drifts. This takes drama out of design research, we think. It would be seen in humble rather than radical terms. We do not need to make brilliant breakthroughs to justify our work; instead, we can improve quality in lots of places, and judge quality by close attention to how we experiment. This we take to be the basic message of new experimentalism to design researchers.

Above we have pointed to how experiments in themselves embed and elicit knowledge and how new experimentalism as a line of thinking in philosophy of science may help us see and value such contributions. Learning from new experimentalism also shows how design objects are not theories in themselves, but a site of intentional drifting.

Chapter 9
At the Frontiers of Knowledge

In this chapter we recap what we have presented so far and point to future directions. As noted in the first lines of Chap. 1, this book is about how constructive design researchers build knowledge through acts of designing. The descriptions and methodologies provided in this book represent our attempt to give legitimacy to a cherished design practice we recognize as *intentional drifting*. Drifting is typical to design, and it cannot be avoided in it. It is better to know it so that it can be turned into a useful instrument of research. We have argued that drifting is a characteristic of constructive design research just as it is a characteristic of design practice, but that there are significant differences as well. In our view, drifting consists of several ill-understood practices that are, however, crucially important to understand how design researchers create knowledge.

We have described a shift to knowledge in design in Chap. 1, and explicated four epistemic traditions in Chap. 3. From there we went on to study how these traditions shape hypothesis construction in design, and presented a Knowledge-Relevance model in Chap. 4. The purpose of the model is to help the constructive design researcher keep track of relation between activities, concerns and contributions of the work. In Chap. 5 we detailed how drifting happens in design experimentation, and in Chap. 6 how evaluation is shaped by epistemic traditions. Finally, in Chap. 7, we elaborated our concept by studying it in terms of the Lars Hallnäs's and Johan Redström's concept 'hermeneutic gap' between research and concept in design.

Key to the presented models and studies are how a constructive design research process can bridge knowledge and relevance — the K-R model — at various stages of research, and how these contribute to traversing the hermeneutic gap. As a research field, constructive design research sits on two chairs: design researchers want to make sure they are accountable — or at least intelligible — to practitioners, but their practice also builds on knowledge and contributes to it, so it must be intelligible to research audiences as well.

As the book shows practices of design do contribute to the building of knowledge, and its practices are distinct. Does this qualify constructive design research as a discipline on its own rights? Recently, Redström (2017) has convincingly argued

P. G. Krogh, I. Koskinen, *Drifting by Intention*, Design Research Foundations,
https://doi.org/10.1007/978-3-030-37896-7_9

that constructive design is able to not only establish and maintain a research discourse it is also as other disciplines capable of challenging programs and occasionally even paradigms. If this is the case, this chapter goes to the future and asks what lies ahead if the field gets more stable?

9.1 Knowledge-Relevance: Talking to Two Audiences

The first thing to note is that juggling the interests of two audiences is easier said than done, and there are few patent solutions to it. As we saw in Chaps. 2 and 3, a leading interaction designer Bill Gaver once distinguished epistemic accountability from aesthetic, and criticized sharply the idea that design should be turned into a science. In our later work, we claimed that his argument is certainly valid, but also too narrow to pay respect to design (Koskinen and Krogh 2015). The reason we expanded his notion from aesthetic accountability to design accountability was that design in a professional context, however, is accountable to not just aesthetic ideals, but to many other types of demands coming from the design world.

Looking at the K-R model at the heart of this book does not provide patent solutions to this question either, but it gives a tool to understand better the demands of the audiences. Our suggestion is to take a look at epistemic tradition around research, and analyze its beliefs before deciding how to deal with multiple audiences. For scientific and engineering audiences, drifting is an ill to be remedied because it does not fit the model of epistemic accountability as Gaver has described it. For a design practitioner, it is not a scientific artefact to be controlled, but an essential part of the craft — and by implication, design research. For those who locate knowledge creation primarily in research discourse, it happens dominantly in debate around design and research, while the actual research methods may vary from almost artistic to rigorous scientific methods. For participatory designers, drifting is something that must happen if the community under study so decides: drifting becomes a function of a community process. The ground premise we have been interested in, therefore, has been the idea that the solution to the two-audience problem in the K-R model is knowing the knowledge tradition in which a research works.

9.2 Epistemic Traditions in Their Full Depth

Epistemic traditions are more than abstractions. They consist of many types of elements that are loosely connected. Following Koskinen et al. (2011), we can distinguish four different types of claims in them: designs; frameworks; theoretical background; and philosophical background. As he claims with his colleagues, saying that constructive design research is only about building things is a misunderstanding because it focuses all too much attention to design objects. They are a necessary component of research, but not sufficient as such, however. Only focusing on them would miss the theoretical and philosophical backgrounds that are equally important in turning an object of design into a research statement. Constructive design research is for us a discipline grounded in artefacts, but were this discipline only about artefacts, we believe it would disappear quick and become a form of professional design practice.

A good case study of how the conceptual scaffolding works comes from Eindhoven, the Netherlands. We have referred to several theses from there, including Wensveen (2005), Frens (2006), Ross (2008), and Deckers (2013). (We have also referred to theses by Djajaginingrat (1998) and Hummels (2000), but like Wensveen, they were done in the Technical University of Delft before these researchers moved to Eindhoven). At one level, all these dissertation included the building of something: Wensveen an alarm clock, Frens a camera, Ross a set of lights, and Deckers experimental prototypes to study phenomenological concepts.

To take Wensveen's and Frens's studies as an example, both were successful in several ways even though they were never meant to go into production (Philips was interested in Frens's work, but he wanted to keep it in the public domain). Yet, design research audiences all over the world know Wensveen's and Frens's prototypes and keep referring to them. At some level, they have become exemplars which is about the highest honor a researcher can get. At another level, both dissertations also produced a framework that became the main research contribution: Wensveen's was 'an interaction frogger,' Frens's 'rich interaction,' and so on. In design debates, people also routinely refer to these frameworks, and they are used as teaching cases in design programs all over the world.

This, however, is where their contribution ends in many ways. Both were well aware of the theoretical framework behind the work they were doing, and referred to it. It was J. J. Gibson's ecological psychology, which came from their supervisor Kees Overbeeke's studies in KU Leuven in Belgium and from his studies into ecological perception with Gerda Smeets in Delft. Wensveen and Frens both referred to Gibson's work, just as they referred to Overbeeke, Smets, and Don Norman, who had popularized Gibson's theories in his book *The Psychology of Everyday Things* (1988). After placing the work into the Gibsonian psychology that located perception into our body rather than in the mind, they quickly move to discuss their design frameworks, however, and did not make claims about contributing to psychology as such. Their work was targeted at design, not psychology.

There was also an even more abstract conceptual layer in their work, which was Gestalt and phenomenological psychology, which Overbeeke had learned in his early years in Leuven. In Leuven, Overbeeke learned about phenomenology through the work of Albert Michotte, who had studied Gestalt psychology in Germany and was well connected to Paris where phenomenology took a psychological turn in Merleau-Ponty's early work. Wensveen and Frens, however, did not refer to this background, nor did other researchers in Eindhoven. It remained in the background until the inaugural lecture of Kees Overbeeke in 2007. In this lecture, he showed the path from artefacts all the way to pragmatism and phenomenology, which had shaped his thinking, and also Gibson's thinking. In the next phase of the program, Overbeeke hired Pierre Levy, a French engineer from a technical university in

Compiegne, France, and worked with the phenomenological philosopher Charles Lenay, who was working in the same school. The first thesis that came out from the phenomenological turn of the Eindhoven program was Deckers (2013).

Wensveen, Frens and Deckers had no illusions about contributing to philosophical discourse. Their focus was on its implications to design. Yet, even though it stayed in the background, it did shape their thinking in other ways. Coming from phenomenological philosophy meant that they did not share many assumptions with, say, Herbert Simon's behaviorism. Instead, they found natural allies in Bill Gaver who had got interested in French Situationism, Italian semiotics (including Ambra Trotto, whose thesis *Rights through Making* was supervised by Overbeeke until his death), and Nordic design, which tended to build on non-Cartesian forms of philosophy. At a deep philosophical level, this work was quintessentially European, which made it incompatible with, say, rational theories of decision-making, American behaviorism, and British utilitarianism.

At the surface level, one main message from this case study is that the conceptual scaffolding has richness that sets limits to what designers can do. It also gives them resources to dig deeper into their work should they want to. At a deeper level, it shows that epistemic traditions, as we call them, are stubborn because they have underlying depth that may go back centuries. In practically any discipline worth its salt, the key focus is on research work (analogous to 'designs' in, Fig. 9.1) and debate about it (analogy is 'frameworks' in Fig. 9.1), and in more advanced levels, theories at the bottom of the discipline. This is maybe the final lesson the Eindhoven program teaches to us. There is no need to be deep in every study; it may be better to leave the depth to places like Overbeeke's inaugural lecture.

9.3 Drifting in Conceptual Scaffoldings

Conceptual scaffoldings are essential elements in drifting, and although they may lead to frustrations, they are also a part of the joy of the discipline as an intellectual endeavour. The scaffolding becomes the site for drifting. It generates perspectives like Eva Deckers's phenomenology (2013), Martin Ludvigsen's Goffmanian interactionism (2006), Katja Battarbee's co-experience (2004), and Tom Djajaginingrat's ecological psychology (1998). It also gives tools to formulate research questions which helps researchers to generate a distance to professional design practice. It is worth noting that these four researchers all are in industry today. One of the questions their work poses is the relationship between design work and the conceptual scaffolding. How can this relationship be defined? Both parts are obviously relevant: Deckers could not have done her studies into perceptual crossing without her prototypes. Yet, thousands of design students create interactive prototypes like hers every year all over the world. The difference between these classroom creations and Deckers's study lies in the scaffolding that specified how her prototypes were meant to be experienced. It informed her construction process, her experiments, and the way in which she drew her conclusions from these experiments. Her prototypes

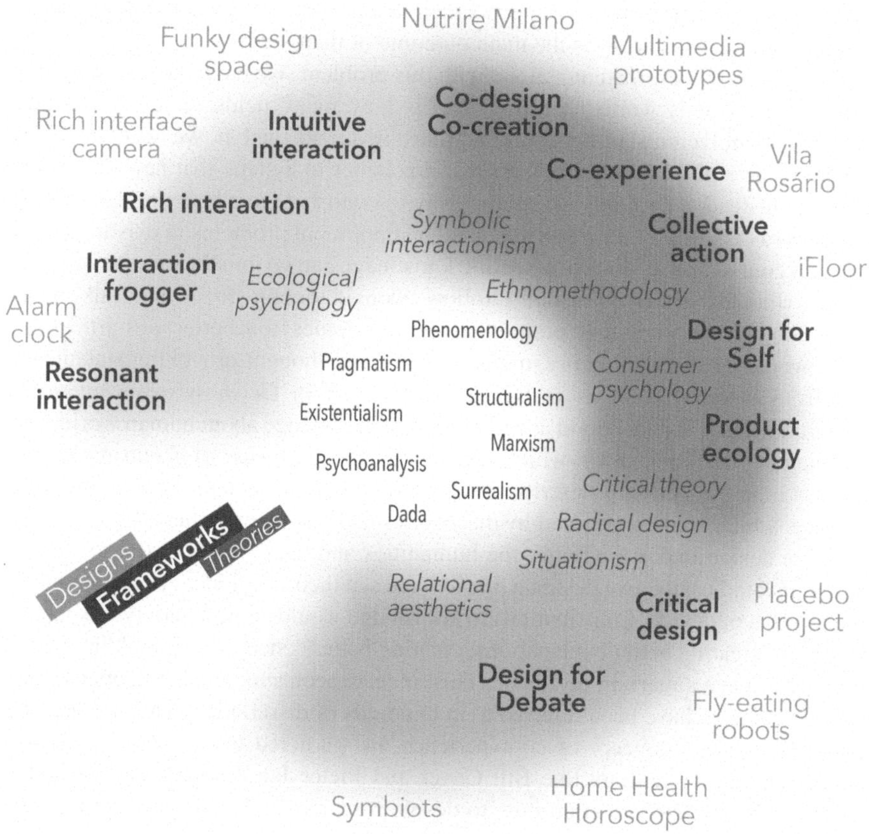

Fig. 9.1 The conceptual scaffolding of constructive design research (Koskinen et al. 2011)

were a necessary part of the larger argument she was making, and they were not meant to stand on their own as research contributions.

When research is done with design objects, they have a certain degree of autonomy, however. Design has to come before it can be studied, and most authors (for example, Stappers 2007; Koskinen et al. 2011; Gaver 2012) very clearly say that these designs usually owe their existence to design practice more than to research practice, which only puts them in motion. When deciding materials, shapes, or when specifying interactions into fractions of the second to make sure the desired experience is there so that it can be studied, it is better to trust design practice. As Carroll and Kellogg (1989) noted, the thing to be studied must be there before it can be studied. This view poses problems at the textbook level of methodology, which usually tells that theory has to specify the experiment: if theory is adjusted after the experiment, theory cannot be wrong. But in research, nature (or second nature, i.e. the humans) has to have a chance to show that theory was wrong. Reversing this order by starting from making, constructive design research in fact give priority to design and tells researchers to explicate their theory afterwards when they are developing frameworks. Theory is the main outcome of their study.

Designers, however, are not alone with this problem. We have said this before in this book, but it is worth repeating: this is how many fields of research argue. Nobody manufactures more Etruscan ruins in Lazio, and if we want to study Shakespeare's plays of Schiller's poems, it is better to assume that new data is not coming. Instead, a perfectly acceptable way to contribute to a better understanding of the bards is to study their texts in detail and point out problems in existing literature. Of course, many kinds of scientific knowledge can go into the process of explication (though designers usually follow Donald Schön 1983 and talk about reflection), but its core is this: progress in research comes from better well-articulated frameworks, changes in theory, methodologies and thought-provoking statements, not from discoveries. This is typical to research in *Geisteswissenschaften*, in research when it is understood as acquisition of knowledge about human world. The German word *Wissenschaft*, which sees research as a process of acquiring knowledge is a better model to describe design research than the modern English word science, which tends to proceed by discoveries and testing experiments.

Were design understood after the humanities and many fields of the social sciences, it could easily give design a priority and see theoretical ideas as an attempt to make sense of design. Contributions to knowledge would come a variety of sources, including creating better explanations, creating better concepts, and building prototypes that leave a mark in discourse. This, indeed, seems to be what happens. Joep Frens's cameras have been referred to in hundreds of dissertations and articles, and Katja Battarbee's concept of co-experience has garnered thousands of citations. Most citations by authors like Bill Gaver and Pieter Jan Stappers are methodic. There are many ways to contribute to design research, and those conceptual scaffoldings researchers create are every bit as crucial as the prototypes.

Again, the Knowledge-Relevance model is useful. It tells that design researchers most work through two different worlds. The model pushes design researchers to find ways to make their craft intelligible to practitioners and researchers alike. It contains a promise to turn design research into a real contributor to design knowledge, but also warns that taking this argument too far pushed research back to practice.

9.4 New Experimentalism

Having said this, we need to take a new look at the design artefact that forms the core of design experiment, which is middle of the K-R model. Our way to understand its role builds on Deborah Mayo's new experimentalism, which says that experiments can produce knowledge on their own, not only through theories they are meant to test. In the corpus we have studied, research is currently characterized by three activities: hypothesis construction, design experiment, and evaluation. In building their experiments, designers sketch ideas, combine them, build mock-ups and when constructing prototypes, typically create several models before deciding which one works best — experiments as trails of potential futures. As we have shown, this is not the case: each breaks into several main types. For instance, we described five ways in which experimentation typically proceeds.

Obviously, experimentation of this sort is closer to design practice than experimental research in science. It does not have a strict methodology, and even if it has, this methodology is not abstract, not grounded in stochastic processes, nor is it procedural in the sense that there are right and wrong ways of doing experiments. Instead, if we are correct, constructive design research can be seen as a form of experimentation with forms, materials, spaces, mechanisms, and programs. The only way to know these things, crucial to design, is to try and err. In line with the promotion of the theory unbound experiment promoted in 'new experimentalism' many of these experiments come from the needs and interests of design rather than theory.

Design experimentation may take several forms, however. Following the taxonomy of the 2011 Koskinen et al. book it can follow a laboratory-style format in which variations of the same design are treated as hypotheses, and evaluation studies determine which one offers the best solution. When done in this way, research tends to use statistical techniques and it tries to achieve a degree of control over the circumstances in which the study is done. Equally often, however, research builds on field research in which design objects are offered to people to be followed in real life where it is impossible to achieve any degree of real control over background variables and their interactions. In critical design practice, design objects can also be taken to conversational encounters with the public, and there is no attempt to gain any degree of control over who is present in these encounters, and who gets to talk.

As this variation shows, experimentation in constructive design research can be done in many ways, but the 2011 book did not say a lot about experiments as such. This makes Mayo's insight useful: it helps us to look at how the actual experiments were made, and how they generate knowledge. A good example is *iFloor*, a prototype at the center of Martin Ludvigsen's *Designing for Social Interaction* (2006) and Andreas Lykke-Olesen's *Space as Interface* (2006). The project built a prototype of an interactive floor that could be used to browse the collections and discuss with other users of Aarhus city library. At the functional level, the prototype is easy to describe. A projector installed to the ceiling projected several icons and text bubbles to the floor at the entrance of the library. When people walked over it, their movement was tracked. When several people gathered around the floor, the cursor moved to areas with most people, so negotiation had to take place in order to intentionally move the cursor. When designing the floor, Ludvigsen and his colleagues had to do lots of design decisions that went far beyond this functionality as such. They could not touch the building, for instance and they had to make sure the floor was easy to clean. These are not typical interaction design concerns, but they are crucially important drivers for design. Their research process for certain had a design component that was by and large almost independent from theoretical concerns, like Ludvigsen's interest in 'collective action' over the floor. The experiment did generate various types of design knowledge, but it was reported in conferences only in part (Krogh et al. 2004). The residue of knowledge is what made the floor work, however. It also made writing conceptual papers about things like pragmatist aesthetics possible. Yet, like the technical work that goes into scientific experiments, it was not reported; still, it was there, enabling the research. In Jung-Joo Lee's (2012) terminology, it formed the interactional substratum that made the design possible in the first place.

9.5 Towards Technical Humanism

Some of the features above make constructive design research sometimes difficult to understand and convey. For example, for hard-core scientists, the idea that reflections can lead to a discipline is hard to accept. In standard texts of methodology, the purpose of research is usually to add increments to knowledge. This happens through careful scholarship that identifies yet unanswered questions and then builds a research plan to examine what would be the answer to this question.

As we have already pointed out, the idea that reflection can build up towards a discipline is closer to some field in the humanities. Carroll and Kellogg (1989) noted a long time ago that the thing must proceed the theory of the thing, such as the mouse arriving before the usability research showing that it functioned as a great pointing device. This is no different from studies of ancient texts or ruins. The best any researcher can do to make sense of Cicero's political writings is to get thoroughly familiar with them thoroughly familiar with scholarship about Ancient Rome during the fall of the Republic. The first two sets of knowledge are necessary for any contribution, but it is the third set that defines what is new. Scholarship on poets long gone progresses through explication, and progress takes the form of a better explication of arguments, facts that are already known, and approaches yet unexplored. It may come from scientific analyses. It is possible to think that a genetic analysis of a sample of DNA in Petrarch's bone could reveal a disease hitherto unknown to scholars, but if the purpose of analysis is to understand his poetry, this argument would lead to obvious and perfectly legitimate objections: how would knowledge of, say, his illnesses, change the meaning of his syntax of rhymes? Would it have to? Possibly, but it is also possible to see value in his art without his personal history.

Research in this picture becomes a process in which researchers build hypotheses from data as they go, and test these hypotheses against new pieces of data. In the humanities, this process would be called the hermeneutical circle (Gadamer 1975), while social scientists usually talk about analytic induction (Koskinen 2003).

Gentes (2017) has recently argued that this is how not only design, but engineering proceeds. She says that engineering has many similarities to the humanities that proceed through explication. She writes:

> But the humanities are not solely considered as an iconographic data exploration that selects a family of objects, analyses what these objects mean and how that meaning is cre-

> ated, they also help with the invention. In fact, the book posits that the humanities have a way of looking at creative practices that are useful to invention. Naming, narrating, composing, debating are all practices that have been thoroughly studied by the humanities. (Gentes 2017: 13–14).

Indeed, there are many fields of design in which novelty comes from 'naming, narrating, composing, [and] debating' rather than through radical innovations. This is how furniture and fashion design practically always work. As Koskinen has noted in an essay about analysis in design,

> Many design disciplines also function like this. The world is full of chairs, but occasionally, better chairs show up. Contemporary designers build on their knowledge of other chairs, but somehow manage to go deeper than the giants from the past. Making a piece of furniture better than Eames, Ponti, Wegner or Kukkapuro is very difficult, but not impossible. (Koskinen 2015: 222).

An improved understanding becomes the key contribution. Previous designs are treated as arguments in on-going conversation rather than as indisputable facts. As Koskinen also notes, many fields of learning proceed through explication: Goethe does not write anymore, but Goethe scholarship does improve over time. This is also the case in fields like clinical medicine that feeds observations into the medical science; case studies in business literature; most historical studies; in ethnographic studies; and so on. Studies with a small number of cases based on existing data are the bread and butter of many disciplines, and they are valuable also in theoretically driven disciplines.

If we follow this line farther, we can also see that their argument opens doors to many kinds of wisdom that the radical scientific version of design research would devalue. If design research is like research in the humanities, anything that sensitizes design researchers to unthought-of gaps in existing arguments may be useful e.g. critical advice on mixing interior design and computational objects (Krogh et al. 2017). It may be a theoretical argument, but it can equally well be a new way of cheaply 3D printing joints in a chair (Magrisso et al. 2018), a novel bending technique of hardwood (like in Alvar Aalto's furniture), or a way to break glass internally so that bacteria can grow in it without getting access to the outside (a typical design school exercise at Master's level).

By expanding inspiration sources, this vision opens design research to many types of sensibilities, many of which are essential to design. By rejecting rationalistic and empiricist philosophies, many researchers have rejected the radical Enlightenment idea that locates progress in reason and gives the mind priority over the body. These philosophical ideas explicitly break with the mind-body dualism that characterizes a good deal of Western philosophy in the sciences, and the idea that progress is rational. Instead, it puts to the fore concepts like Michael Polanyi's (2009) tacit knowledge and Martin Heidegger's (1963) notion of being-in-the-world, which was a part of a movement away from empiricist and rationalist philosophies that put the mind over the body.

9.6 The Limits of Eurocentricism

These observations lead to our second last point. If it is so that when the philosophical basis of thinking changes in European discourse, the way of understanding knowledge changes too. There is a reason to believe that it also will change when it is built on non-Western patterns of thinking. A few times, we have noted that most of the work we have used to build up our message come from Northern Europe and North America, research cultures on constructive design research here are very specific. They are considered high-tech countries with a preference on technologically complicated and multidisciplinary research for highly specialized knowledge. The research tradition in design is driven by writing and priority is given to generally applicable results and less to the 'ultimate particulars' (as Nelson and Stolterman describe) that design as discipline deliver. These conditions may differ from region to region: like any other human activity, research is embedded in its surrounding society, thus other conditions and ideals may govern research in other environments. It may be stating the obvious that constructive design research needs to stay open to this, but looking to the dominant production in field, it is less trivial.

If we start to treat the research included in this book as a model for the rest of the world, we may be propounding views that are ill-fitted outside is region of origin. At worst research turns into a form of colonialism and even re-colonialism, occasionally self-inflected. The problem is structural rather than individual. Most design research communities have their origin in Europe and concentrate their review practices to European hands, communities, and fora. Its base is particular, but the temptation, within its current research fora to say its practices are universally valid, is obvious.

If we look to the historic tales of design applied in industry first, it is clear that it has seen many universalizing tendencies over the years. Its histories are usually Western, and many of the leading schools and museums have espoused universal criteria over taste, technology, and techniques — the impact of the Ulm school on the current smart products distributed globally is evident. Most theories that have entered disciplines of design have been formal and again Western, including semiotics, information theory, and many strands of user-centered design, and Silicon Valley's business models are in global scale, offering solutions for the planet regardless of location. Industry, however, knows that sometimes design proposals has to come from local circumstances. Architecture, interior design and graphic design have their regionalisms (car styling in the fifties in America had nothing to do with European modernist ideals) and also global players, but they also have strong voices telling that all good design is ultimately local. We must be wary of believing that industrial success proves that thinking behind design is valid. This line of argument confounds power with validity. If older forms of colonialism built on military power that helped the colonial powers to import raw materials to industrial centres, many of these current universalizing tendencies take an economic form. Even pro bono design suffers from colonialism. The Global South in particular has seen lots of one-shot projects done to organize photo-ops rather than to help people, and bringing in

Michelin starred chefs to teach people in slums in Africa how to cook is not only silly, but also arrogant.

Design researchers are not free of these assumptions either. Unlike their peers in industry, researchers have more freedom to choose their assumptions, including those about how to position themselves on the debate between universal and particular aspects of design. These choices maybe mostly ideological in research, but they are still enormously consequential in terms of their implications to research practice. In particular, the last 20 years have seen several psychological theories being imported into design as if they were universal in nature. In Europe, these tendencies are evident mostly in the Netherlands, where many leading researchers build on psychological theories that are thought to be valid all over the world. Yet, the universality of these theories is questionable. The Cambridge University Press paper (Henrich et al. 2010) *The weirdest people in the world?* Show that more than 2/3 of the participants in psychology studies came from the United States, and 96% come from Western industrialized countries. To add to the bias, two-thirds of the American subjects were undergraduate students in psychology.

If we deconstruct the universal ideals a bit, we can find several cracks in the system and see how they help us to understand alternative realities. First, design may not be nowhere near as universal as it seems. Italian design sells badly in Scandinavia, and Scandinavian design is clearly out of place in the popular markets in France and Russia. When Scandinavian ideals were introduced to Barcelona in the sixties, they carried messages from free and equal Europe against Franco's dictatorship through their distinctive, non-Castilian aesthetic (Viviana Narotzky 2000). Design ideals that look universal in textbooks build in many ways on cultural and political ideals.

Another way to approach this question is to leave Europe and North America and see how design research is done in other parts of the world. Brazil, may be a good example of the contour of design in Latin America. As Adèlia Borges (2011) has argued, design in Brazil in particular was defined in European terms from an optimistic, rationalistic era of the fifties and the sixties. It built on semiotics, sociology and information theory, and New Typography, all presenting design ideals in universal ways. This discourse has dominated academic writing about design in Brazil, just as it has dominated its exhibitions and media images. As Borges notes, its hidden design culture, however, built on craft and it is mostly vernacular and highly sensitive to the local circumstances in that vast country. The craftsmen of tropical Brazil, for instance, use natural rubber, nuts and woods in ways consistent with indigenous traditions, but they also renew it. The gap between these two worlds in Brazil is a good example of the danger of treating Western-world practices as universally valid. Similar stories can be heard in East Africa, which has good design schools, but little industry to provide customers for designers. Design in European sense is an alien implant in places like Kenya and Uganda.

Still another question is whether the universalistic premises work even in places where they came from, and whether they will work there. Many universalistic premises in European design were a reaction to production technology. For example, rectangular and perpendicular forms are relatively easy from a manufacturing standpoint. Not to mention what the classical tools for drawing (triangles, circular tools, parallel ruler etc.) have meant for shaping our physical world; the explosion of ellipsoids and amorphous shapes coincides with the spreading of computers as design office tool. Smart manufacturing and 3D printing technologies are currently expanding these forms. These tools of manufacturing in turn may break the market-driven imperative coming from economies of scale. When corporations aim at global markets, their design tends to become what Dunne and Raby (2001) memorably called Hollywood blockbusters — meant for everyone to the extent that they become bland and essentially uninteresting. Micro-production and limited design may be an economically viable future based on local 4.0 tools of fabrication and up-cycling.

9.7 Constructive Design Research Beyond Europe

Design research has particularistic tendencies as well. At one end are recent craft revivals in Italy, Scandinavia and the UK, which has pushed design into a fiercely local, and even folkloristic direction (Summatavet 2005). Research that had gone into this has usually been decidedly local, meant to work in local context. Artefacts and services take their form from local community, and business models are embedded into the local community (the main difficulty to designers is to give up that part of their ego, which says they are experts and hence have a right to knowledge). A common thread in all these cases is a shift in philosophical thinking behind design, which has tended to get more respectful towards particulars.

Where does constructive design research sit on this field? As we have seen, it puts design into the core of research and sees it as the central instrument of knowledge creation. Making design is to a high degree dependent on local circumstances, including materials, machinery, and the social organization of work. Glassblowers in Murano have their secrets. This is also true of research. The word knowledge, on the other hand, suggests abstracting away from these particulars. It may be defined in universal terms, but also in particular terms: it is impossible to understand research in the Designing the Quality of Interaction group in Eindhoven without Kees Overbeeke and his inner circle of students. Thought patterns, reading, and connections that animated this group's work are unique. Yet, as we have seen, its foundational ethos was to generalize from its studies towards a goal, which is a mathematical model of interaction. What could be more universalizing than this?

On the other hand, Bill Gaver and his co-authors have strongly emphasized that their work is meant to produce design objects, not theory and not methods. The ethos is deeply particularizing. A quick assessment, then, can go to both directions. Constructive design research has been a part of the trend that has questioned universalism by valuing local knowledge, but it plays the game of research, which is almost by necessity universalistic in its ambition. This nascent debate remains unsolved.

If we turn to constructive design research directly, we can find some cues from our corpus. Practically all methods in the corpus come from European world. For example, contextual inquiry builds on anthropology and sociology and cultural probes build on Situationism. This is true of theories and philosophical ideas too. With the exception of pragmatism, an American outgrowth of Hegel and Kant, its philosophy is almost fully European. Frameworks like relational aesthetics, 'hacktivism', collective action, and rich interaction all build on European thinking. They all are constitutive to research and its main outcomes, but European to the core.

This is true even when we turn to those PhD students who come from outside Europe. For example, Karthikeya Acharya (2016) roots his thinking on Kenneth Frampton's (patronizing) views on critical regionalism (1992). Andrea Botero (2013) and Joanna Saad-Sulonen (2014) built on activity theory, which builds on Russian psychology and builds on Marx. The list could easily be extended.

Looking at this subset of dissertations, however, we see few possible cases that come from outside the West. Most importantly, Andrea Jucide (2014) built her thinking on Paulo Freire's pedagogy, which of course builds on Marx, Lukacs and Sartre, but also on local circumstances in poor parts of Brazil. What they got from Freire was a very fine design ethos: Freire's philosophy acts as a powerful reminder that designers have to be on the side of the underdog, but not help them to become new masters. That would defeat the purpose of the pedagogy, which is to help people to free themselves from self-made (and structurally grounded) prisons. If one Master-Slave dialectic is turned around, it is yet another Master-Slave dialectic. His pedagogy was also a precursor to co-design. Following Freire, Judice built research methods and constructive methods on local meanings, resources and skills, and treated local 'health agents' who fought tuberculosis in the Vila Rosario district of Rio de Janeiro as their teachers and co-equals, avoiding lecturing and taking expert roles. Interestingly for our inquiry, this position was a result of a close reading of the Swedish designer Pelle Ehn's thesis *Work-Oriented Design of Computer Artifacts* (1988), by now a classic chronicle of the philosophical aspects of participatory design and two of its constitutive projects. In his book, Ehn had an important footnote that that told how Freire had been an important source of ideas to young researchers and computer scientists in the early days of participatory design.

These non-European aspects of constructive design research, however, are still ill-understood and would require another study. Yet, they show that there are ways to do research that do not fit the patterns developed in the West. Design can be done in local ways. If this is the case, these ways may open up alternative ways of seeing Western design traditions too, as in the case of Freire and participatory design.

9.8 The Future of Design as a Knowledge-Based Discipline

We are strong believers in constructive design research, and claim that there is enough evidence to show that it has been and is useful in the design world. While design research in the eighties tended to be equal to art history and around year 2000 turned to user studies, current research scene is much more complicated. It is also much better able to respond to needs coming from design practice and from design students. It has opened doors to other disciplines by giving design researchers a language that helps them to make themselves intelligible in research world. It also works well with industry. Constructive design research has played a significant role in this transformation by bringing research much closer to design experience.

Perhaps most importantly, we believe it works well in universities too. One of the issues shaping design at the moment is that design has become increasingly academic, but design practice has followed its own demands. This is not new. Design has a long history of failed attempts at getting into research. The attempts have usually failed perhaps because research is shaped by criteria coming from research funding bodies and disciplines outside design promoting their ideals. Talking about The Department of Design Research at London's RCA, McIntyre writes:

> A list of studies carried out between 1962 and 1980 reveals a number of interesting changes in emphasis in the research base. In the early 1960s, all of the projects resulted in products. Ten out of fifteen of the major projects were carried out on a contract basis… In the 1970s the study of design methods was replaced by the study of principles for erecting and manipulating models of the things or systems being designed. The focus on the cognitive processes of design activity completely replaced the development of real products.
>
> The change from commercial projects to theoretical enquiries was, in part, a pragmatic decision. The academic side of their work was, at the time, funded by the Science Research Council and by charitable research funding agencies. The studies of method in live projects gave way to the study of the principles for erecting and manipulating models of the things being designed, one step further away from the things.
>
> But the studies of real design projects undertaken had to appear to the funding body to be academically respectable: the Department had to give up many ideas for projects on the grounds that they would have judged as too commercially motivated. (McIntyre 1995: 61).

McIntyre's analysis tells about one department only, but can also serve as a useful reminder of a tension tearing any research field that tries to be practical. Over time, research programs have a tendency to become internally focused. Just like practitioners, researchers develop a core set of beliefs that are not negotiable, and their way of working is shaped by the need to get research grants. Incentive systems in universities (they are the main site of research) add to this pressure. In fact, this has been happening in HCI rapidly over the last ten years or so. A formerly fairly practical field focusing on usability and user experience has become increasingly a place in which people create academic careers. This has shifted the focus of the community away from what is industrially useful to increasingly imaginative and artistic directions.

Constructive design research is not free from a tendency to get academic, but it has been a useful corrective to it. It has given young designers a way to do design research in ways that do not create a large gap to professional practice. By doing so, it has also helped to maintain design skills in academic work, and it has given a rise to a generation of professors who can still talk to practitioners because their methods are, if not exactly the same, though at least mutually intelligible and in the service of a generative process that culminates with an artefact. It has also extended the skill set of practice by bringing methods from professional practice into a research environment. In this shift, designers have been able to experiment with methods that their peers in reality would not dare to try.

There are pitfalls in this shift, of course. The main danger is going too fast too far into theoretical and methodological abstractions that are too far from the experience of design practitioners. Christopher Alexander, an early proponents of design research who later turned against methodology, warns about the perils of abstraction through an example with an obvious sarcastic overtone:

> Something that is relatively sensible becomes extremely absurd when you call it methodology. Murray Silverstein and I have been designing a building. We have been… putting wood stakes in the ground to indicate the organization of the ground as we create it. A friend….jokingly… said we should write this up and send it into Emerging Methods in Environmental Design and Planning. We really laughed…. Anything that is actually real is scorned by methodologists. Anything that is legitimate methodology is accepted precisely because it is so remote from everyday flesh and blood. (Alexander, quoted in McIntyre 1995: 61–62).

Alexander's sarcasm serves as a good warning for research in the future. Constructive design research cannot avoid the tension between design and being methodical. Its fate in the long run may be to become yet another failed attempt to turn design, elusive as it is, into a putative science. Yet, at least so far, it has been a useful addition to design schools and professional practice, and it has managed to give a sandbox in which designers can play with concepts, processes, and methods they would not dare to try in business.

This is the nature of being on the frontier. Usually it offers only risks and leads to either failures or modest successes. Sometimes, however, the risk pays off and researchers hit gold. This has happened in constructive design research a few times, and those new practices it has generated have become reality in universities. They also have become a reality in design practice in some of the most experimental and historically complicated markets of the world. Design practitioners can indeed sell contextual studies, experience prototypes, and cultural probes to their customers. We believe that by keeping in mind the history of failed attempts to root design in science, and by holding ourselves accountable not only to the academic world but also to professional design practice, we may indeed be able to continue keeping their work relevant to both worlds. Our final word is a plea for modesty, however: there will be failures on the pathways to the frontier.

Our final word takes us back to the concept of drifting. In rally driving, drifting is what happens when you do sharp turns at high speed and the rear end of the car skids in a controlled fashion. We see this as the characteristic of drifting in constructive design research. It should be respected rather than taken for granted, and it should be acknowledged as a crucial resource in design research rather than something to be paced out. If it is lost from research, research may lose wisdom that exists in professional practice, either current or new. Should this happen, it has to face the full force of Alexander's sarcasm about research.

Chapter 10
Corpus

The corpus we have been using in this book consists of PhD dissertations done in the constructive vein. It has been divided into primary and secondary categories. The reason lies in the history of the book. It started from a series of papers by the first author. For those papers, his primary corpus was mostly Scandinavian, but he had added a few cases from other countries that had shaped work.

When the second author joined, the authors expanded the corpus, and started to test the claims in the original set of papers against the expanded corpus. It expanded the scope of the book from design experimentation to knowledge (the main theme in Chaps. 1, 2 and 3) and to the relationship between design and conceptual scaffolding (Chaps. 7–8) and the geographical limits of the book (Chap. 9).

The word corpus is mainly familiar in the humanities and linguistics, but it is much less well-known outside the French context in design. Our use of the term builds on Annie Gentes, who defines corpus in the following manner:

> Semiotics scholars define a corpus as a group of texts (linguistic and/or visual) that belong to the same genre.[1] Texts of the same genre share structural and pragmatic characteristics in so far as they are not only similar in style, narrative structure, or visual organization, but also as they are discussed by social actors as targeting the same audiences, for the same purposes, with similar textual strategies. A corpus therefore is based on formal properties that are generally considered similar by a group of actors. For instance, science fiction authors, editors, critics, and readers can agree that certain novels belong to this group of literature based on formal and experiential qualities… (Gentes 2017: 90).

She also notes that this is a formal definition and distances herself from it by noting that it is ultimately the audience who decides what belongs to a corpus. Also for her, no corpus is exhaustive; it is a tool for answering a research question.

> Beyond the formal properties of the examples here, one must therefore understand that they are generally considered as belonging to the same group of intellectual productions by the actors. Nevertheless, the constitution of the corpus does not pretend to be exhaustive. It is always but a segment of the actual production. The purpose is to gather enough material to

[1] Rastier, 2001.

P. G. Krogh, I. Koskinen, *Drifting by Intention*, Design Research Foundations,
https://doi.org/10.1007/978-3-030-37896-7_10

> answer a research question. The method therefore consists in defining a corpus of textual elements and considering what are the shared elements and their purpose in the text and in the experience of their readers, spectators, and users. To do so, the humanities researcher can and will base her selection following the cue given by the actors. But she can also add new elements and compare productions with the hypothesis that they offer similar aesthetic or social experiences. (Gentes 2017: 90–91).

We think this is a perfectly appropriate way to understand our collection of theses. Like in pragmatic linguistics focusing on talk, in which the language of "sampling" tends to falter, and it would be plainly silly to say that the data is perfect, we do not claim our corpus is perfect. Also, it evolves quick.

Yet, it consists of a few well-known bodies of research coming mostly from North-Western Europe. This helps to constitute the stylistic unity Gentes talks about. The corpus from Jutland, Denmark, and Helsinki, Finland is quite systematic: it comes from Danish research in debt to participatory design, and Finnish research in debt to user-centered and empathic design. In many cases, we have built around the two main sites Aarhus in Denmark and Helsinki in Finland, though, so for instance on community design, we have built on work done in Helsinki (for instance, Miettinen 2007; Isohanni 2006; Judice, A. 2014; Judice, M. 2014; Garcia 2017). The selection is not a statement of any sort: it mostly reflects the fact that the authors know the background stories from Jutland and Helsinki, and both have built a broad, forward-looking portfolio over the years.

Other dissertations are selected to fill in the gaps and raise questions. These include dissertations like Joep Frens's (2006) by now canonical work, and Kristina Niedderer's (2004) dissertation paved the way for craft research community in the UK. Frens's work, for instance, comes from a well-known and influential body of Dutch research that saw its beginnings in Delft in the nineties, and found its culmination in the Dutch Research Council's research program about research through design in 2016. British work finds unity from Frayling's concept of research through design, and it has animated a good deal of design research in former polytechnic universities in a craft and HCI, and often in their cross-over (as in the case of Wallace et al. 2013, 2018).

The word 'corpus' of course, has obvious roots in Latin, but it is not commonly used in design literature and requires a word of explanation. We use it the same way linguists in pragmatics use it. It is a body we study to learn about the wider space out there. This outer space in linguistics is not manageable by typical sampling techniques, so the only way to manage it is to develop a large enough corpus to go through it as Sam Spade, the private eye in The Maltese Falcon written by Dashiell Hammett. In essence, we followed leads in the corpus until we came up with the four epistemic traditions that seemed to be robust and wide enough to cover what we were seeing in our exploration. By no means is our analysis perfect and final; its value lies in the word drifting, in the way we have shed light on it by identifying our four traditions, and in the ways in which we have studied research practices through these traditions.

10.1 Primary Corpus

Acharya, K. (2016). *Opening the electrome*. Helsinki: Aalto University.

Bang, A. L. (2011). *Emotional value of applied textiles*. Kolding: Kolding School of Design.

von Busch, O. (2008). *Fashion-able: Hacktivism and engaged fashion design*. Göteborg: School of Design and Crafts (HDK), Faculty of Fine, Applied and Performing Arts, University of Gothenburg.

Deckers, E. (2013). *Perceptive qualities in systems of interactive products*. Eindhoven: Technische Universiteit Eindhoven.

Dindler, C. (2010). *Fictional space in participatory design of engaging interactive environments*. Aarhus University.

Dunne, A. (1999). *Hertzian Tales: Electronic products, aesthetic experience, and critical design*. London: RCA.

Fogtmann, M. H. (2011). *Designing with the body in mind: Kinesthetic, empathy, interaction*. Arkitektskolen Aarhus.

Frens, J. W. (2006). *Designing for rich interaction: Integrating form, interaction, and function*. Eindhoven: Technische Universiteit Eindhoven.

Kinch, S. (2014). *Designing for atmospheric experiences — Taking an architectural approach to interaction design*. Aarhus School of Architecture.

Krag, M. M. S. (2017). *Transformation on abandonment: A new critical practice?* Aarhus: Aarhus School of Architecture.

Kurvinen, E. (2007). *Prototyping social action*. Helsinki: University of Art and Design.

Lee, J.-J. (2012). Against method. In *The portability of method in human-centered design*. Helsinki: Aalto University.

Ludvigsen, M. (2006). *Designing for social interaction: Physical, co-located social computing*. Aarhus: Aarhus School of Architecture.

Lynggaard, A. B. (2012). *Homing interactions:-tactics and concepts for highly mobile people*. Aarhus: Aarhus School of Architecture.

Mäkelä, M. (2003). *Saveen piirtyvia muistoja*. Helsinki: Aalto University. ['*Memories Written in Clay*', in Finnish].

Mattelmäki, T. (2006). *Design probes*. Helsinki: University of Art and Design.

Niedderer, K. (2004). *Designing the performative object: A study in designing mindful interaction through artefacts*. Plymouth: University of Plymouth.

Rasmussen, M. K. (2015). *Changing the shape of interaction: Shape-changing interfaces*. Aarhus: Aarhus University, Department of Computer Science.

Ross, P. (2008). *Ethics and aesthetics in intelligent product and system design*. Eindhoven: Technische Universiteit Eindhoven.

Summatavet, K. (2005). *Folk tradition and artistic inspiration*. Helsinki: University of Art and Design.

Trotto, A. (2011). *Rights through making: Skills for pervasive ethics*. Eindhoven: Technische Universiteit Eindhoven.

Wallace, J. (2007). *Emotionally charged: A practice-centred enquiry of digital jewellery and personal emotional significance*. Sheffield: Shefield-Hallam University.

Wensveen, S. (2004). *A tangibility approach to affective interaction*. Eindhoven: Technische Universiteit Eindhoven.

Worbin, L. (2010). *Designing dynamic textile patterns*. Borås: University of Borås.

Wu, Y. (2017). *Bicycles and plants*. Helsinki: Aalto University.

10.2 Expanded Corpus

Aakko, M. (2016). *Fashion in between*. Helsinki: Aalto University.

Ahde-Deal, P. (2013). *Women and jewelry*. Helsinki: Aalto University.

Auger, J. (2012). *Why robot? Speculative design, the domestication of technology and the considered future*. London: RCA.

Bakker, S. (2013). *Design for peripheral interaction*. Eindhoven: Technische Universiteit Eindhoven.

Battarbee, K. (2004). *Co-experience: Understanding user experiences in social interaction*. Helsinki: University of Art and Design Helsinki.

Berg, A. (2014). *Artistic research in public space*. Helsinki: Aalto University.

Botero, A. (2013). *Expanding design spaces. Design in communal endeavors*. Helsinki: Aalto University.

Bruns, M. (2010). *Relax: Inherent feedback during product interactions to reduce stress*. Eindhoven: Technische Universiteit Eindhoven.

Hummels, C. (2000). Gestural design tools: Prototypes. In *Experiments and scenarios*. Delft: Delft University of Technology.

Djajaginingrat, T. (1998). *Cubby: What you see is where you act*. Delft: Delft University of Technology.

Hengeveld, B. (2011). *Designing linguaBytes. A tangible language learning system for non- or hardly speaking toddlers*. Eindhoven: Technische Universiteit Eindhoven.

Garcia, C. G. (2017). *Design as freedom*. Helsinki: Aalto University.

Grondahl, L. (2004). *Experiences in theatrical space*. Helsinki: University of Art and Design.

Groth, C. (2017). *Making sense through hands*. Helsinki: Aalto University.

Isohanni, T. (2006). *Arabia, Arabia: taiteellinen toiminta osana asuinympäristön suunnittelua, tapaus Arabianranta, Helsinki*. Helsinki: University of Art and Design. *'Arabia, Arabia: Artistic Activity in Planning a Housing District,* Case Arabianranta, Helsinki, *Finland'*, in Finnish.

Jacucci, G. (2004). *Interaction as performance*. Oulu: University of Oulu.

Jucide, A. (2014). *Design for hope*. Helsinki: Aalto University.

Judice, M. (2014). *You are important!* Helsinki: Aalto University.

Kabiito, R. (2014). *Meaning-making in visual culture: The case of integrating Ganda indigenous knowledge with contemporary art practice in Uganda*. Helsinki: Aalto University.

Keller, I. (2005). *For inspiration only*. Delft: Delft University of Technology.

Kohtala, C. (2016). *Making sustainability*. Helsinki: Aalto University.

Kuusk, K. (2016). *Crafting sustainable smart textile services*. Eindhoven: Technische Universiteit Eindhoven.

Kymalainen, T. (2015). *Science fiction prototypes as design outcome of research*. Helsinki: Aalto University.

Lankoski, P. (2010). *Character-driven game design*. Helsinki: Aalto University.

Lucero, A. (2009). *Co-designing interactive spaces with and for designers*. Eindhoven: Technische Universiteit Eindhoven.

Marttila, T. (2018). *Platforms of co-creation*. Helsinki: Aalto University.

Miettinen, S. (2007). *Designing the creative tourism experience*. Helsinki: University of Art and Design.

Nimkulrat, N. (2009). *Paperness: Expressive material in textile art from an artist's viewpoint*. Helsinki: University of Art and Design.

Paavilainen, H. (2013). *Dwelling with design*. Helsinki: Aalto University.

Raijmakers, B. (2007). *Design documentaries. Using documentary film to inspire design*. London: RCA.

Saad-Sulonen, J. (2014). *Combining participations. Expanding the locus of participatory E-planning by combining participatory approaches in the design of digital technology and in urban planning*. Helsinki: Aalto University.

Säde, S. (2001). *Cardboard mock-ups and conversations*. Helsinki: University of Art and Design.

Salgado, M. (2009). *Designing for an open Museum*. Helsinki: University of Art and Design.

Scheible, J. (2010). *Empowering mobile art practice*. Helsinki: Aalto University.

Siikamäki, R. (2006). *Glass can be recycled forever utilisation of end-of-life cathode ray tube glasses in ceramic and glass industry*. Helsinki: University of Art and Design.

Sleeswijk-Visser, F. (2009). *Bringing the everyday life of people into design*. Delft: Delft University of Technology.

Soini, K. (2015). *Toward resident-oriented housing renovation with collaborative design*. Helsinki: Aalto University.

Tromp, N. (2013). *Social Design*. Delft: Delft University of Technology.

Vaajakallio, K. (2012). *Design games as a tool, a mindset and a structure*. Helsinki: Aalto University.

Vines, J. (2011). *Ageing futures: Towards cognitively inclusive digital media productions*. Plymouth: University of Plymouth.

Ylirisku, S. (2013). *Frame it simple!* Helsinki: Aalto University.

Literature

Acharya, K. (2016). *Opening the electrome*. Helsinki: Aalto University.

Adriaan Ianus, K. (2005). For Inspiration Only: Designer Interaction with Informal Collections of Visual Material. [s.n.].

Ahde-Deal, P. (2013). *Women and jewelry: A social approach to wearing and possessing Jewelry*. Helsinki: Aalto University.

Aicher, O. (2009). The Bauhaus and Ulm. In P. Oswalt (Ed.), *Bauhaus conflicts, 1919–2009: Controversies and counterparts* (pp. 174–186). Ostfildern: Hatje Cantz.

Alexander, C. (1971). The state of the art in design methods. *DMG Newsletter, 5*(3), 3–7.

Archer, B. (1995). The nature of research. *Co-Design Journal, 2*, 6–13.

Archer, L. B. (1979). Design as a discipline. *Design Studies, 1*, 17–20.

Bang, A. L. (2011). *Emotional value of applied textiles*. Kolding: Kolding School of Design.

Bang, A. L., & Eriksen, M. A. (2014). Experiments all the way in programmatic design research. *Proceedings of Nordic Design Research Conference*, 2013. https://doi.org/10.14434/artifact.v3i2.3976.

Bang, A. L., Peter, K. G., Ludvigsen, M., & Markussen, T. (2012). The role of hypothesis in constructive design research. In *Proceedings of the art of research 2012*. Helsinki: Aalto University.

Bass, J., & Kirkham, P. (2011). *Saul Bass: A life in film and design*. London: Laurence King Publishing.

Battarbee, K. (2004). *Co-experience: Understanding user experiences in social interaction*. Helsinki: University of Art and Design Helsinki.

Battarbee, K., & Koskinen, I. (2005). Co-Experience: User Experience as Interaction. CoDesign 1: 5–18. https://doi.org/10.1080/15710880412331289917.

Baudrillard, J. (1994). *Simulacra and simulation*. Ann Arbor: University of Michigan Press.

Beaver, J., Kerridge, T., & Pennington, S. (2009). *Material beliefs*. London: Goldsmiths, Interaction Research Studio.

Bertola, P. (Ed.). (2009). *Sistema design Milano / Milan design system*. Milano: Abitare Segesta.

Bierut, M. (2007). *Seventy-nine short essays on design*. New York: Princeton Architectural Press.

Biggs, M. (2000). The foundations of practice-based research: Introduction. In *Working papers in art and design 1*. Hertfordshire: University of Hertfordshire. http://www.herts.ac.uk/artdes1/research/papers/wpades/vol1/vol1intro.html.

Binder, T., De Michelis, G., Ehn, P., Linde, P., Jacucci, G., & Wagner, I. (2011). *Design things*. Cambridge: MIT Press.

Binder, T., & Redström, J. (2006, November 1–4). Exemplary design research. *DRS Wonderground Conference*, Lisbon.

P. G. Krogh, I. Koskinen, *Drifting by Intention*, Design Research Foundations,
https://doi.org/10.1007/978-3-030-37896-7

Blessing, L., & Chakrabarti, A. (2009). *DRM, a design research methodology*. London: Springer.

Blomberg, J., Giacomi, J., Mosher, A., & Swenton-Wall, P. (1993). Ethnographic field methods and their relation to design. In D. Schuler & A. Namioka (Eds.), *Participatory design: Principles and practices* (pp. 123–155). Boca Raton: CRC Press.

Bossen, C., Dindler, C., & Iversen, O. S. (2010). User gains and PD aims: Assessment from a participatory design project. In *Proceedings of participatory design conference* (pp. 141–150). New York: ACM.

Bødker, S., & Kyng, M. (2018). Participatory design that matters — Facing the big issues. *ACM Transaction on Computer-Human Interaction, 25*(1), 1–31.

Bødker, S., Kyng, M., Ehn, P., Kammersgaard, J., & Sundblad, Y. (1987). *A utopian experience – on design of powerful computer-based tools for skilled graphic workers*. Aldershot: Gower Publishing.

Borges, A. (2011). *Design+craft: The Brazilian path*. Sao Paulo: Terceiro Nome.

Borgmann, A. (1995). The depth of design. In R. Buchanan & V. Margolin (Eds.), *Discovering design* (pp. 13–22). Chicago: University of Chicago Press.

Botero, A. (2013). *Expanding design spaces. Design in communal endeavors*. Helsinki: Aalto University.

Bowers, J. (2012). The logic of annotated portfolios: Communicating the value of research through design. In *Proceedings of the designing interactive systems conference* (pp. 68–77). New York: ACM.

Box, G. E. P. (1979). Robustness in the strategy of scientific model building. In R. L. Launer & G. N. Wilkinson (Eds.), *Robustness in statistics* (pp. 201–236). New York: Academic.

Box, G. E. P. (1976). Science and statistics. *Journal of the American Statistical Association, 71*, 791–799.

Brandt, E. (2006). Designing exploratory design games: A framework for participation in participatory design? In *Proceedings of the ninth conference on participatory design expanding boundaries in design*. New York: ACM Press.

Brandt, E., & Binder, T. (2007). Experimental design research: Genealogy, intervention, argument. *Proceedings of International association of societies of design research*, Hong Kong.

Branzi, A. (2010). Design come Professione Di Massa. *Disegno Industriale, 7*(42/43).

Branzi, A. (2013). Seven degrees of separation. In *The New Italian design* (pp. 14–17). Milano: Triennale di Milano.

Buchanan, R. (2001). Design research and the new learning. *Design Issues, 17*, 3–23.

Bunge, M. (1977). *Treatise on basic philosophy: Ontology I: The furniture of the world*. London: Springer.

von Busch, O. (2008). *Fashion-able: Hacktivism and engaged fashion design*. Göteborg: School of Design and Crafts (HDK), Faculty of Fine, Applied and Performing Arts, University of Gothenburg.

Carroll, J., & Kellogg, W. A. (1989). Artifact as theory-Nexus: Hermeneutics meets theory-based design. In *Proceedings of the SIGCHI conference on human factors in computing systems* (pp. 7–14). New York: ACM.

Cash, P., Stanković, T., & Štorga, M. (2014). Using visual information analysis to explore complex patterns in the activity of designers. *Design Studies, 35*, 1–28.

Collins, H. M. (1985). *Changing order: Replication and induction in scientific practice*. Beverly Hills/London: Sage.

Cross, N. (1984). *Developments in design methodology*. New York: Wiley.

Cross, N. (1995). *Discovering design ability*. Chicago: University of Chicago Press.

Cross, N. (1999). Design research: A disciplined conversation. *Design Issues, 15*, 5–10.

Cross, N. (2006). *Designerly ways of knowing*. London: Springer.

Cross, N. (2018). Developing Design as a Discipline. *Journal of Engineering Design, 29*, 691–708. https://doi.org/10.1080/09544828.2018.1537481.

Dalsgaard, P. (2014). Pragmatism and design thinking. *International Journal of Design, 8*, 143–155.

Dalsgaard, P. (2010). Research in and through design. An interaction design research approach. In *Proceedings of the 22nd conference of the computer-human interaction special interest group of Australia on computer-human interaction* (pp. 200–203). New York: ACM.
Deckers, E., Hummels, C., Feijs, L., & Wensveen, S. (2013). *Perceptive qualities in systems of interactive products*. Eindhoven: Technische Universiteit Eindhoven. http://repository.tue.nl/753907.
Desmet, P. (2002). *Designing emotions*. Delft: Delft University.
Dindler, C. (2010). *Fictional space in participatory design of engaging interactive environments*. Aarhus: Aarhus University.
Dindler, C., Iversen, O. S., & Krogh, P. G. (2011). Engagement through mixed modalities. *Interactions, 18*(4), 34–39.
DiSalvo, C. (2012). *Adversarial design*. Cambridge: MIT Press.
Djajadiningrat, J. (1988). *CUBBY: What you see is where you act*. Delft: Delft University, the Netherlands.
Djajadiningrat, T., Matthews, B., & Stienstra, M. (2007). Easy Doesn't Do It: Skill and expression in tangible aesthetics. *Personal and Ubiquitous Computing, 11*(8), 657–676.
Dorst, K. (2015). *Frame innovation: Create new thinking by design*. Cambridge: MIT Press.
Dreyfus, H. (1972). *What computer still can't do: A critique of artificial reason*. Cambridge: MIT Press.
Dunne, A. (2006). *Hertzian Tales: Electronic products, aesthetic experience, and critical design*. Cambridge: MIT Press.
Dunne, A., & Raby, F. (2001). *Design noir: The secret life of electronic objects*. Basel: August/Birkhäuser.
Dunne, A., & Raby, F. (2013). *Speculative everything: Design, fiction, and social dreaming*. Cambridge: MIT Press.
Durbin, P. (Ed.). (1991). *Critical perspectives on nonacademic science and engineering*. Bethlehem: Lehigh University Press.
Ehn, P., & Kyng, M. (1991). Cardboard computers: Mocking-it-up or hands-on the future. In J. Greenbaum & M. Kyng (Eds.), *Design at work: Cooperative design of computer systems* (pp. 169–195). Hillsdale: Lawrence Erlbaum Associates.
Ehn, P. (1988). *Work-oriented design of computer artifacts*. Stockholm: Arbetslivscentrum.
Emerson, R. M., & Pollner, M. (1989). On the uses of members responses to researchers accounts. *Human Organization, 47*, 189–198.
Fallman, D. (2007). Why research-oriented design Isn't design-oriented research. On the tensions between design and research in an implicit design discipline. *Knowledge, Technology & Policy, 20*, 193–200.
Feldman, R. (2003). *Epistemology*. Upper Saddle River: Prentice Hall.
Fogtmann, M. H. (2011). *Designing with the body in mind: Kinesthetic, empathy, interaction*. Aarhus: Arkitektskolen Aarhus.
Fogtmann, M. H., & Kinch, S. (2011). The design researcher in quandary: The conflicting roles of being both a designer and a design researcher. In *Proceedings of the second conference on creativity and innovation in design DESIRE* (Vol. 11, pp. 193–196). New York: ACM.
Forlizzi, J., & Battarbee, K. (2004). Understanding experience in interactive systems. In *Proceedings of designing interactive systems* (pp. 261–268). New York: ACM.
Frampton, K. (1992). *Modern architecture: A critical history*. London: Thames and Hudson.
Frankel, L., & Racine, M. (2010). The complex field of research: For design, through design, and about design. In *Design research society conference design and complexity*. Montreal: Design Research Society.
Frayling, C. (1993). Research in art and design. In *Royal college of art research papers* (Vol. 1, pp. 1–5). London: RCA. http://researchonline.rca.ac.uk/384/.
French, W. L., & Bell, C. H. (1999). *Organization development: Behavioral science interventions for organization improvement* (6th ed.). New Jersey: Prentice Hall.
Frens, J. (2006). *Designing for rich interaction: Integrating form, interaction, and function*. Eindhoven: Technische Universiteit Eindhoven.

Froukje, S. V. (2009). *Bringing the everyday life of people into design*. Delft: Delft University of Technology.

Gadamer, H.-G. (1975). *Truth and method*. London: Sheed & Ward.

Gamper, M. (2010). *100 chairs in 100 days and its 100 ways*. London: Dent-De-Leone.

Garcia, C. G. (2017). *Design as Freedom*. Helsinki: Aalto University.

Gaver, W. (2002, November). Presentation about cultural probes. Powerpoint presentation, Helsinki. http://smart.uiah.fi/luotain/pdf/probes-seminar/GaverPROBES.pdf/. Accessed 18 April 2009.

Gaver, W., & Bowers, J. (2012). Annotated portfolios. *Interactions, 19*(4), 40–49.

Gaver, W. (2012). What should we expect from research through design? In *Proceedings of the SIGCHI conference on human factors in computing systems* (pp. 937–946). New York: ACM.

Gaver, W. W., Hooker, B., and Dunne, A. (2001). The Presence Project. RCA Computer Related Design Research, London.

Presence Project, Gaver, W., Hooker, B., Dunne, A., & Farrington, P. (2001). *CRD projects series*. London: RCA.

Geertz, C. (1973). *The interpretation of cultures*. New York: Basic Books.

Gentes, A. (2017). *The In-discipline of design*. Dordrecht: Springer.

Gibson, J. J. (1979). *The ecological approach to visual perception*. New York: Taylor & Francis.

Goffman, E. (1978). *The presentation of self in everyday life*. Harmondsworth: Penguin.

Goldschmidt, G. (1992). Criteria for design evaluation: A process-oriented paradigm. In Y. E. Kalay (Ed.), *Evaluating and predicting design performance* (pp. 67–75). New York: Wiley.

Goldschmidt, G., & Tatsa, D. (2005). How good are good ideas? Correlates of design creativity. *Design Studies, 26*, 593–611.

Habermas, J. (2015). *Knowledge and human interests*. New York: Wiley.

Hallnäs, L., & Redström, J. (2006). *Interaction design: Foundations, experiments*. Borås: University of Borås.

Hallnäs, L., & Redström, J. (2001). Slow technology — Designing for reflection. *Personal and Ubiquitous Computing, 5*, 201–212.

Hamilton, J., & Jaaniste, L. O. (2009). The effective and the evocative: Practice-led research approaches across art and design. In R. Woodrow (Ed.), *Interventions in the public domain*. Brisbane: ACUADS Publishing, Queensland College of Art, Griffith University.

Härkäsalmi, T. (2008). *Runkokuituja lyhytkuitumenetelmin: kohti pellavan ja hampun ympäristö-myötäistä tuotteistamista*. Taideteollinen korkeakoulu: Helsinki.

Hasdell, P. (2016). Participatory design: Re-evaluation as a socio-material assembly. In *Proceedings of Ethnographic Praxis in industry conference* (pp. 313–326).

Hatchuel, A., & Weil, B. (2008). Entre concepts et connaissances: éléments d'une théorie de la conception. In A. H. e. B. Weil (Ed.), *Les Nouveaux Régimes de la Conception: Langages, Théories, Métiers*. Paris: Vuibert/Cerisy.

Hatchuel, A., & Weil, B. (2011). *Experts in organizations: A knowledge-based perspective on organizational change*. Berlin: De Gruyter.

Hauser, S., Oogjes, D., Wakkary, R., & Verbeek, P.-P. (2018). An annotated portfolio on doing Postphenomenology through research products. In *Proceedings of designing interactive systems conference* (pp. 459–471). New York: ACM.

Hassenzahl, M. (2004). The interplay of beauty, goodness, and usability in interactive products. *Human-Computer Interaction, 19*, 319–349.

Henrich, J., Heine, S. J., & Norenzayan, A. (2010). The weirdest people in the world? *Behavioral and Brain Sciences, 33*, 61–83.

Heiskanen, I. (1976). How to skin an eel with too sharp a knife and without proper sand on hand (or personal experiences in applying the results of organizational research). *Interfaces, 6*, 93–97.

Hevner, A. R., March, S. T., Park, J., & Ram, S. (2004). Design science in information systems research. *Management Information Systems Quarterly, 28*, 75–105.

Hummels, C. (2000). Gestural design tools: Prototypes. In *Experiments and scenarios*. Delft: Delft University of Technology.

Husserl, E. (1965). *Phenomenology and the crisis of philosophy*. New York: Harper and Row.
Ihde, D. (2006). The designer fallacy and technological imagination. In J. R. Dakers (Ed.), *Defining technological literacy* (pp. 121–131). New York: Palgrave Macmillan.
Isohanni, T. (2006). *Arabia, Arabia: taiteellinen toiminta osana asuinympäristön suunnittelua, tapaus Arabianranta, Helsinki*. Helsinki: University of Art and Design. *'Arabia, Arabia: Artistic Activity in Planning a Housing District, Case Arabianranta, Helsinki, Finland'*, in Finnish.
Jacobs, J. (1961). *The death and life of great American cities*. New York: Random House.
Jacucci, G. (2005). *Interaction as performance*. Oulu: University of Oulu.
Jones, C. (1991). *Designing designing*. London: Architecture Design and Technology Press.
Jones, J. C. (1992). *Design methods*. New York: Wiley.
Jucide, A. (2014). *Design for hope*. Helsinki: Aalto University.
Kasanen, E., Lukka, K., & Siitonen, A. (1993). The constructive approach in management accounting research. *Journal of Management Accounting Research, 5*, 241–246.
Kensing, F., & Blomberg, J. (1988). Participatory design: Issues and concerns. *Computer Supported Cooperative Work, 7*(3–4), 167–185.
Kettunen, I. (2013). *Mielekkyyden muotoilu*. Kuusamo: Aatepaja.
Kinch, S. (2014). *Designing for atmospheric experiences — Taking an architectural approach to interaction design*. Aarhus: Aarhus School of Architecture.
Koskinen, I. (2003). Empathic Design in Methodic Terms. In I. Koskinen, K. Battarbee, & T. Mattelmäki (Eds.), *Empathic design*. Helsinki: IT Press.
Koskinen, I. (2016). Agonistic, Convivial, and Conceptual Aesthetics in New Social Design. *Design Issues, 32*: 18–29. https://doi.org/10.1162/DESI_a_00396.
Koskinen, I., Mikkonen, J., Ahde, P., Eckoldt, K., Helgason, T., Hänninen, R., Jiang, J., Niskanen, T., & Schultz, B. (2009). Hacking a car: Re-embodying the design classroom. In *Proceedings of Nordic design research conference*. Available at http://nordes.org.
Koskinen, I., Zimmerman, J., Binder, T., Redström, J., & Wensveen, S. (2011). *Design research through practice. From the lab, field, and showroom*. Waltham: Morgan Kaufmann.
Koskinen, I. (2015). Four cultures of analysis in design research. In P. A. Rodgers & J. Yee (Eds.), *The Routledge companion to design research*. London: Routledge.
Koskinen, I., & Krogh, P. G. (2015). Design accountability: When design research entangles theory and practice. *International Journal of Design, 9*, 121–127.
Koskinen, I., & Frens, J. (2017). Research Prototypes. *Archives of Design Research, 30*, 5–14. https://doi.org/10.15187/adr.2017.08.30.3.5.
Krogh, P. G., Markussen, T., & Bang, A. L. (2015). Ways of drifting—Five methods of experimentation in research through design. In *Proceedings of ICoRD15 — Research into design across boundaries* (pp. 39–50). New Delhi: Springer.
Krogh, P. G., Petersen, M. G., O'Hara, K., & Groenbaek, J. E. (2017). Sensitizing concepts for socio-spatial literacy in HCI. In *Proceedings of conference on human factors in computing systems* (pp. 6449–6460). New York: ACM.
Krogh, P., Martin, L., & Lykke-Olesen, A. (2004). Help Me Pull That Cursor' A Collaborative Interactive Floor Enhancing Community Interaction. Australasian Journal of Information Systems 11. https://doi.org/10.3127/ajis.v11i2.126.
Kuhn, T. (1962). *The structure of scientific revolutions*. Chicago: University of Chicago Press.
Kurvinen, E. (2007). *Prototyping social action*. Helsinki: University of Art and Design Helsinki.
Kuutti, K. (1996). Activity theory as a potential framework for human-computer interaction research. In B. A. Nardi (Ed.), *Context and consciousness. Activity theory and human-computer interaction*. Cambridge: MIT Press.
Lakatos, I. (1970). Falsification and the methodology of scientific research programmes. In *Criticism and the growth of knowledge* (pp. 91–196). New York: Cambridge University Press.
Lakatos, I. (1974). The role of crucial experiments in science. *Studies in History and Philosophy of Science, 4*, 309–325.

Latour, B. (2004). How to talk about the body? The normative dimension of science studies. *Body and Society, 10*, 205–229.
Latour, B. (1987). *Science in action: How to follow scientists and engineers through society*. Cambridge: Harvard University Press.
Latour, B., & Woolgar, S. (2013). *Laboratory life: The construction of scientific facts*. Princeton: Princeton University Press.
Lawson, B., & Dorst, K. (2013). *Design expertise*. London: Routledge.
Lee, J.-J. (2012). *Against method: The portability of method in human-centered design*. Helsinki: Aalto University.
Lewin, K. (1946). Action research and minority problems. *Journal of Social Issues, 2*, 34–46.
Ludvigsen, M. (2006). *Designing for social interaction: Physical, co-located social computing*. Aarhus: Aarhus School of Architecture.
Lynch, M. (1993). *Scientific practice and ordinary action: Ethnomethodology and social studies of science*. Cambridge, UK: Cambridge University Press.
Lynggaard, A. B. (2012). *Homing interactions: Tactics and concepts for highly mobile people*. Aarhus: Aarhus School of Architecture.
Magrisso, S., Mizrahi, M., & Zoran, A. (2018). Digital joinery for hybrid carpentry. In *Proceedings of conference on human factors in computing systems*. New York: ACM.
Mäkelä, M. (2003). *Saveen piirtyvia muistoja*. Helsinki: Aalto University. ['Memories Written in Clay', in Finnish].
Mäkelä, M., & Routarinne, S. (2006). *The art of research: Research practices in art and design*. Helsinki: University of Art and Design Helsinki.
Maldonado, T. (1972). *Design, nature and revolution: Toward a critical ecology* (trans: Domandi, M.). New York: Harper and Row.
Maldonado, T. (1984) Ulm Revisited. *Rassegna*, Anno 19/3, settembre. (trans: Sparado, F.).
Manzini, E. (2015). *Design, when everybody designs: An introduction to design for social innovation*. Cambridge: MIT Press.
Manzini, E. (2016). Design culture and dialogic design. *Design Issues, 32*, 52–59.
Markussen, T., & Bang, A. L. (2012). Dynamic research sketching. A new explanatory tool for understanding theory construction in design research. In *Proceedings of the 8th international design research society conference*. Thailand: Chulalongkorn University Bangkok.
Markussen, T., Krogh, P. G., & Bang, A. L. (2015). On what grounds?: An intra-disciplinary account of evaluation in research through design. In *Proceedings of 6th international association of societies of design research conference* (pp. 1415–1429). Brisbane.
Mattelmäki, T. (2006). *Design probes*. Aalto University.
Mattelmäki, T., Kirsikka, V., & Ilpo, K. (2013). What happened to empathic design? Design Issues, 30(1), 67–77. https://doi.org/10.1162/DESI_a_00249.
Mattelmäki, T., & Visser, F. S. (2011). Lost in Co-X: Interpretations of co-design and co-creation. In *Proceedings of the 4th conference on design research* (pp. 1–12).
Mayo, D. (1996). *Error and the growth of experimental knowledge*. Chicago: University of Chicago Press.
McCullough, M. (2005). *Digital ground: Architecture, pervasive computing, and environmental knowing*. Cambridge: MIT Press.
McIntyre, J. (1995). The department of design research at the royal college of art: Its origins and legacy 1959-1988. In C. Frayling (Ed.), *Design of the times: One hundred years of the royal college of art* (pp. 58–62). Shepton Beauchamp: Somerset.
Mendini, A., Weiss, P., & Nollert, A. (2012). *Alessandro Mendini*. Nürnberg: Verlag für moderne Kunst.
Meroni, A. (2017). *Creative communities: People inventing sustainable ways of living*. Milan: Edizioni Polidesign.
Meroni, A., & Sangiorgi, D. (Eds.). (2011). *Design for services*. Adelshot: Gower Publishing.
Miettinen, S. (2007). *Designing the Creative Tourism Experience*. Helsinki: University of Art and Design.
Mikkonen, J. (2016). *Prototyping interactions*. Tampere: Tampere Univrsity of Technology.

Mitcham, C. (1986). Computers: From Ethos and Ethics to Mythos and Religion. Notes on the new frontier between computers and philosophy. *Technology in Society, 8*, 171–201.
Mitchell, W. J. (2000). *E-Topia: 'Urban life, Jim — But not as we know it*. Cambridge: MIT Press.
Mogensen, P. (1992). Towards a provotyping approach in systems development. *Scandinavian Journal of Information Systems, 4*, 31–53.
Moggridge, B. (2007). *Designing Interactions*. Cambridge: MIT Press.
Moran, D. (2002). *Introduction to phenomenology*. London: Routledge.
Narotzky, V. (2000). A different and new refinement' design in Barcelona, 1960–1990. *Journal of Design History, 13*, 227–243.
Nelson, H. G., & Stolterman, E. (2003). *The design way: Intentional change in an unpredictable world*. Cambridge: MIT Press.
Niedderer, K. (2004). *Designing the performative object: A study in designing mindful interaction through artefacts*. Plymouth: University of Plymouth.
Niedderer, K., & Roworth-Stokes, S. (2007). The role and use of creative practice in research and its contribution to knowledge. In *Proceedings of IASDR*. Hong Kong SAR.
Niinimäki, K. (2011). *From disposable to sustainable: The complex interplay between design and consumption of textiles and clothing*. Helsinki: Aalto University.
Nowotny, H., Scott, P., Gibbons, M., & Scott, P. B. (2008). *Re-thinking science: Knowledge and the public in an age of uncertainty*. London: Polity.
Odom, W., Wakkary, R., Lim, Y.-k., Desjardins, A., Hengeveld, B., & Banks, R. (2016). From research prototype to research product. In *Proceedings of conference on human factors in computing systems* (pp. 2549–2561). New York: ACM.
O'Neil, C. (2017). *Weapons of math destruction: How big data increases inequality and threatens democracy*. London: Penguin Books.
Overbeeke, K., Wensveen, S., & Hummels, C. (2006). Design research: Generating knowledge thorough doing. In *Proceedings of third symposium of design research* (pp. 51–69). Swiss Design Network: Geneva.
Paavilainen, H. (2013). *Dwelling with design*. Helsinki: Aalto University.
Paul F. M. (1954). The Information Capacity of the Human Motor System in Controlling the Amplitude of Movement. *Journal of Experimental Psychology, 47*, 381–391. https://doi.org/10.1037/h0055392.
Peirce, C. S. (1979). Reviews, correspondence, and bibliography. 3. Print. In C. Hartshorne et al. (Eds.), *Collected papers of Charles Sanders Peirce* (Vol. 8). Cambridge: Belknap Press of Harvard Univ. Press.
Peteri, V. (2017). Bad enough ergonomics: A case study of an office chair. *SAGE Open, 7*(1). https://doi.org/10.1177/2158244016685135.
Petersen, M., Krogh, P. G., Ludvigsen, M., & Lykke-Olesen, A. (2005). Floor interaction HCI reaching new ground. In *Extended abstracts on human factors in computing systems*. New York: ACM.
Petersen, M. G., Iversen, O. S., Krogh, P. G., & Ludvigsen, M. (2004). Aesthetic interaction: A pragmatists aesthetics of interactive systems. In *Proceedings of designing interactive systems* (pp. 269–276). New York: ACM.
Pihl, G. T., Oebro, L. F., Ammentorp, J., & Osther, P. (2018). From active surveillance to active survivor: A pilot study. *European Urology Supplements, 17*(2), e91.
Plowman, L., Rogers, Y., & Ramage, M. (1995). What are workplace studies for? In *Proceedings of European conference on computer-supported cooperative work* (pp. 309–324). Dordrecht: Springer.
Poggenpohl, S. (2009). Time for change: Building a design discipline. In S. Poggenpohl & K. Sato (Eds.), *Design integrations: Research and collaboration* (pp. 3–22). Chicago: University of Chicago Press.
Polanyi, M. (2009). *The tacit dimension*. Chicago: University of Chicago Press.
Popper, K. (2002). *The logic of scientific discovery*. London: Routledge.

Radice, B., & Sottsass, E. (1993). *Ettore Sottsass: A critical biography*. Milan: Rizzoli International Publications.
Rancière, J. (2004). *The politics of aesthetics: The distribution of the sensible*. (trans: Rockhill, G.). London: Continuum.
Rasmussen, M. K. (2015). *Changing the shape of interaction: Shape-changing interfaces*. Aarhus: Department of Computer Science, Aarhus University.
Rasmussen, M. K., Pedersen, E. W., Petersen, M. G., and Hornbæk, K. (2012). Shape-Changing Interfaces: A Review of the Design Space and Open Research Questions. In Proceedings of the 2012 ACM Annual Conference on Human Factors in Computing Systems. USA: ACM Press. https://doi.org/10.1145/2207676.2207781.
Rastier, F. (2001). *Arts et sciences du texte*. Paris: Presses universitaires de France.
Redström, J. (2017). *Making design theory*. Cambridge: MIT Press.
Redström, J. (2011). Some notes on programme-experiment dialectics. In *Proceedings of Nordic design research conference*. http://www.nordes.org/opj/index.php/n13/article/view/91.
Rittel, H. W. J., & Webber, M. M. (1973). Dilemmas in a general theory of planning. *Policy Sciences, 4*, 155–169.
Rogers, Y. (2011). Interaction design gone wild: Striving for wild theory. *Interactions, 18*, 58–62.
Rogers, Y., & Marshall, P. (2017). Research in the wild. *Synthesis Lectures on Human-Centered Informatics, 10*, 1–97. https://doi.org/10.2200/S00764ED1V01Y201703HCI037.
Ross, P. (2008). *Ethics and aesthetics in intelligent product and system design*. Eindhoven: Technische Universiteit Eindhoven.
Rothman, L. (2017). Manhattan Almost Had a Highway Running Across It. Here's What It Would Have Looked Like. Time. Accessed March 11, 2019. http://time.com/4742520/jane-jacobs-robert-moses-drawings/.
RTD Ethos. https://www.researchthroughdesign.org/rtdethos/. Accessed 30 May 2018.
Ruskin, J. (1989). *The seven lamps of architecture*. New York: Dover Publications.
Rust, C., Mottram, J & Till, J. (2007). *AHRC review of practice-led research in art design & architecture*. Arts and Humanities Research Council, Sheffield Hallam University, Nottingham Trent University & University of Sheffield. http://archive.org/details/ReviewOfPractice-ledResearchInArtDesignArchitecture.
Saad-Sulonen, J. (2014). Combining Participations. Expanding the Locus of Participatory E-Planning By Combining Participatory Approaches in the Design of Digital Technology and in Urban Planning. Helsinki: Aalto University.
Säde, S. (2001). Cardboard Mock-ups and Conversations. Helsinki: University of Art and Design.
Sanders, E. B.-N. (2005). Information, Inspiration and Co-Creation. In Proceedings of the 6th International Conference of the European Academy of Design. University of the Arts Bremen.
Sanders, E. B.-N., & Stappers, P. J. (2008). Co-creation and the new landscapes of design. *CoDesign, 4*, 5–18.
Schmid, A.-F. (2018). The philosophical underpinnings of design theory. In P. Vermaas & S. Vial (Eds.), *Advancements in the philosophy of design*. Dordrecht: Springer.
Schön, D. (1983). *The reflective practitione: How professionals think in action*. San Francisco: Jossey-Bass.
Schuler, D., & Namioka, A. (1993). *Participatory design: Principles and practices*. Boca Raton: CRC Press.
Seago, A., & Dunne, A. (1999). New methodologies in art and design research: The object as discourse. *Design Issues, 15*, 11–17.
Sevaldson, B. (2010). Discussions & movements in design research. *FORMakademisk, 3*(1).
Siikamäki, R. (2006). *Glass can be recycled forever: Utilisation of end-of-life cathode ray tube glasses in ceramic and glass industry*. Aalto University.
Simon, H. (1969). *The sciences of the artificial*. Cambridge, MA.
Snow, C. P. (1959). *The two cultures and the scientific revolution*. Cambridge: Cambridge University Press.
Sottsass, E. (2018). *Triennale Exhibition: Sagottanta arredamenti Srl*. https://www.sag80.com/en/news/ettore-sottsass-triennale.html. Accessed 5 July 2018.

Stappers, P. (Jan 2007). Doing design as a part of doing research. In R. Michel (Ed.), *Design research now* (pp. 81–91). Basel: Birkhäuser.

Stappers, P. J., & Giaccardi, E. (2017). The Encyclopedia of Human-Computer Interaction, 2nd Ed. https://www.interaction-design.org/literature/book/the-encyclopedia-of-human-computer-interaction-2nd-ed/research-through-design

Stappers, P. J., Sleeswijk Visser, F., & Keller, I. (2014). The role of prototypes and frameworks for structuring explorations by research through design. In P. Rodgers & J. Yee (Eds.), *Routledge Companion to Design Research* (pp. 163–174). Florence: Routledge., 2014.

Steffen, D. (2013). Characteristics and interferences of experiments in science, the arts and in design research. *Proceedings of Nordic Design Research Conference*, Copenhagen. http://nordes.org.

Steinø, N., & Markussen, T. (2011). Design research between design and research. In J. Dehs, M. W. Esbensen, & C. P. Pedersen (Eds.), *When architects and designers write, draw, build?* Aarhus: Aarhus School of Architecture. 2011.

Stolterman, E. (2008). The nature of design practice and implications for interaction design research. *International Journal of Design, 2*, 55–65.

Suppes, P. (1960). *A comparison of the meaning and uses of models in mathematics and the empirical sciences*. Technical report no. 33. Stanford: Institute for Mathematical Studies in the Social Sciences of Applied Mathematics and Statistical Laboratories.

Summatavet, K. (2005). *Folk tradition and artistic inspiration: A womans life in traditional Estonian jewelry and crafts as told by Anne and Roosi*. Helsinki: University of Art and Design Helsinki.

The Art of Research. (2017). http://artofresearch2017.aalto.fi/previous_events.html. Accessed 30 May 2018.

Thomsen, J. R., Krogh, P. G., Schnedler, J. A., & Linnet, H. (2018). Interactive interior and proxemics thresholds: Empowering participants in sensitive conversations. In *Proceedings of conference on human factors in computing systems*. New York: ACM.

Thomsen, J., & Schnedler, J. (2017). *Designing interactive interiors for value-driven healthcare informed by socio-spatial concerns*. Aarhus: Aarhus University. Master thesis.

Tilley, A., & Dreyfuss, H. (2002). *The measure of man and woman: Human factors in design*. New York: Wiley.

Trotto, A. (2011). *Rights through making: Skills for pervasive ethics*. Eindhoven: Technische Universiteit Eindhoven.

Vaajakallio, K. (2012). *Design games as a tool, a mindset and a structure*. Helsinki: Aalto University.

Valtonen, A. (2007). *Redefining industrial design: Changes in the design practice in Finland*. Helsinki: University of Art and Design Helsinki.

Verganti, R. (2009). *Design-driven innovation: Changing the rules of competition by radically innovating what things mean*. Cambridge: Harvard Business Press.

Wallace, J. (2007). *Emotionally Charged: A Practice-Centred Enquiry of Digital Jewellery and Personal Emotional Significance*. Sheffield: Shefield-Hallam University.

Wallace, J., Rogers, J., Shorter, M., Thomas, P., Skelly, M., & Cook, R. (2018). The SelfReflector: Design, IoT and the high street. In *Proceedings of conference on human factors in computing systems*. New York: ACM. 2018.

Wallace, J., Wright, P. C., McCarthy, J., Green, D. P., Thomas, J., & Olivier, P. (2013). A design-led inquiry into personhood in dementia. In *Proceedings of the SIGCHI conference on human factors in computing systems* (pp. 2617–2626). New York: ACM.

Wasson, C. (2002). Collaborative work: Integrating the roles of ethnographers and designers. In S. Squires & B. Byrne (Eds.), *Creating breakthrough ideas: The collaboration of anthropologists and designers in the product development industry* (pp. 71–91). Westport: Bergin Garvey.

Wensveen, S. (2005). *A tangibility approach to affective interaction*. Delft: Technical University of Delft.

Wilson, F. R. (1999). *The Hand: How its use shapes the brain, language, and human culture*. New York: Vintage.

Wolf, T. V., Rode, J. A., Sussman, J., & Kellogg, W. A. (2006). Dispelling 'design' as the black art of CHI. In *Proceedings of conference on human factors in computing systems* (pp. 521–530). New York: ACM.

Worbin, L. (2010). *Designing dynamic textile patterns*. Borås: University of Borås.

Wu, Y. (2017). Bicycles and Plants. Helsinki: Aalto University.

Ziman, J. (2000). *Real science: What it is and what it means*. Cambridge: Cambridge University Press.

Zimmerman, J., & Forlizzi, J. (2008). The role of design Artifacts in design theory construction. *Art, 2*, 41–45.

Zimmerman, J., Forlizzi, J., & Evenson, S. (2007). Research through design as a method for interaction design research in HCI. In *Proceedings of conference on human factors in computing systems* (pp. 493–502). New York: ACM.

Zimmerman, J., Stolterman, E., & Forlizzi, J. (2010). An analysis and critique of research through design: Towards a formalization of a research approach. In *Proceedings of designing interactive systems* (pp. 310–319). New York: ACM.

Zivanovic, A., Auger, J., & Loizeau, J. (2009). Carnivorous domestic entertainment robots. In *Proceedings of tangible and embedded interaction* (pp. 127–130). New York: ACM Press.

Index

P. G. Krogh, I. Koskinen, *Drifting by Intention*, Design Research Foundations,
https://doi.org/10.1007/978-3-030-37896-7

E

The manufacturer's authorised representative in the EU is Springer Nature Customer Service Centre GmbH, Europaplatz 3, 69115 Heidelberg, Germany. If you have any concerns regarding our products, please contact ProductSafety@springernature.com

Printed and bound by CPI Group (UK) Ltd, Croydon, CR0 4YY
22/09/2026
02221058-0020